AF802846

ADVANCES IN CAD/CAM

CONTENTS

Contributing Authors vii

Preface ix

1. Directions for Engineering Data Exchange for Computer Aided Design and Manufacturing 1
   John J. Clancy

2. Integrating Engineering and Manufacturing 37
   William D. Beeby

3. Applications and Benefits of CAD/CAM Data in the Manufacturing Environment 56
   John H. Schulz

4. Release Control, Configuration Management, Design Retrieval, and Remote Viewing of CAD Generated Product Definition Drawings 72
   Robert E. Perri

5. Automation Technology: The Key to Our Productive Future 77
   R. G. Freeman III

6. Creating an EDP Architecture via Computer Graphics—A Tool for Systems Integration Management 82
   Robert T. Bannon

7. Operation of Jukebox Optical Disk over a Local Area Network 90
   Oliver E. Bessette and Michael E. Sullivan

8. Raster Scan Technology Opens New Frontiers for the Microimaged Engineering Drawing 95
   Gail N. Boyd

9. Computer-Aided Technical Publications 98
   Deborah Gail Brown

10. Image Processing Technology Applied to Engineering Document Management Systems 113
    Rebecca D. Costa

11. Solids Modeling in Engineering Documentation 119
    Russell J. Doty

12. The Impact of Optical Storage Technology on Image Processing Systems 126
    Gerald T. Durbin

13. Mass Storage Systems from the User Point of View 131
    John K. Gingrich

14. 3M Multifunction Printer for CAD 142
    Charles C. Lee and Robert R. Jasper

15. The National Coal Liquefaction Technology Data Base System 148
    George W. Pukanic and William C. Peters

16. Problems in Integrating and Automating a Multi-Site/Multi-Sponsor Test Data Collection and Reporting System 163
    Patrick A. Race

17. Optical Storage of Digital Data on Video Disks for Distributing Engineering Data and Technical Documentation 170
Jane N. Ryland

18. An Integrative Approach to Engineering Data and Automatic Project Coordination 174
Richard L. Segal

19. Strategic Project Methodology for a Computer-Aided Engineering Design System 179
Larry D. Sieck

20. DEMAND: The First Operational Optical Disk-Based Demand Printing System 182
Lee F. Wilbraham

Contributing Authors

Robert T. Bannon	EDP Planning AT&T Communcations	5C 210 Bedminster, NJ 07921
William D. Beeby	William D. Beeby & Associates	P.O. Box 505 Kent, WA 98031
Oliver E. Bessette	RCA	P.O. Box 8 Elmer, NJ 08318
Gail N. Boyd	Versatec, Inc.	2710 Walsh Avenue, MS 3-14 Santa Clara, CA 95051
Deborah Gail Brown	Ford Aerospace & Communication	307 19th Street Huntington Beach, CA 92648
John J. Clancy	McDonnell Douglas Automation Company	P.O. Box 516 St. Louis, MO 63166
Rebecca D. Costa	OMEX	2323 Owen Street Santa Clara, CA 95051
Russell J. Doty	Applicon	32 Second Avenue Burlington, MA 01803
Gerald T. Durbin	Storage Technology Corporation	14008 Maebeth Drive SE Huntsville, AL 35803
R. G. Freeman, III	McDonnell Douglas Corporation	P.O. Box 516 St. Louis, MO 63166
John K. Gingrich	Alpharel, Incorporated	18 Ledgewood Circle Topsfield, MA 01983
Charles C. Lee	Engineering Systems Division 3M Company	235-2F-32, 3M Center St. Paul, MN 55144
Robert E. Perri	CAD/CAM Services Lockheed Missiles & Space Corporation	Bldg. 102, ORGN 19-37 P.O. Box 504 Sunnyvale, CA 94086
William C. Peters	Pittsburgh Energy Tech Center U.S. Department of Energy	P.O. Box 10940 Pittsburgh, PA 15236
George W. Pukanic	Pittsburgh Energy Tech Center U.S. Department of Energy	P.O. box 10904 Pittsburgh, PA 15236
Patrick A. Race	U.S. Army Tank-Automotive Command	Warren, MI 48090
Jane N. Ryland	Reference Technology Inc.	1832 North 55th Street Boulder, CO 80302

John H. Schulz	Manufacturing Support McDonnell Aircraft Company	P.O. Box 516 St. Louis, MO 63166
Richard L. Segal	California Institute of Technology	256-80 Pasadena, CA 91125
Michael E. Sullivan	RCA	Bldg. 13-7-4 Front & Cooper Streets Camden, NJ 08102
Lee F. Wilbraham	Advanced System Concepts Xerox Special Information Systems	P.O. Box 7018 Pasadena, CA 91109

ADVANCES IN CAD/CAM

1

DIRECTIONS FOR ENGINEERING DATA EXCHANGE FOR COMPUTER AIDED DESIGN AND MANUFACTURING

John J. Clancy
Senior Vice President
Industry and Product Management
McDonnell Douglas Automation Company
P.O. Box 516
St. Louis, Missouri 63166

Abstract

During the past 10-15 years, the development and enhancement of CAD/CAM (Computer Aided Design/Computer Aided Manufacturing) Systems has greatly changed the engineering world. Along with the many improvements that have been realized, problems have also arisen. One such problem is that of exchanging product definition data between dissimilar CAD/CAM Systems.

This presentation considers this problem, past methods of dealing with it, the current approaches, and some directions for the future. The history of standardization efforts, including the development of IGES (Initial Graphics Exchange Specification) and its extensions for defining a "neutral" representation format, is discussed. Alternate approaches to addressing data exchange and an indication of some on-going and anticipated activities in this area are presented.

The McDonnell Douglas Corporation (MDC) has played a leading role in many of the activities in this area. Contributions of MDC in these past and on-going efforts as well as some of their current activities are presented here.

Much of what is being done in this area will have a potentially significant impact on the definition and communication of complete product definition data between engineering and manufacturing. The resolution of the problem of product data exchange is important to increasing productivity in the factory of the future.

1. INTRODUCTION

With the rapid development of CAD/CAM (Computer Aided Design/Computer Aided Manufacturing) Systems over the past 10-15 years, many industries have realized the potential for increased productivity by using computerized representations of their products.

2. NEED

In addition to the many improvements and advantages realized by the advancing CAD/CAM technology, some problems also arose. The entry of many CAD/CAM vendors into the marketplace with systems of varying sizes and capabilities, together with the fact that the development of a product usually involves several engineering and manufacturing disciplines, has led to an environment where not only contractors and subcontractors may be using different CAD/CAM Systems, but even within a given corporation several systems are being used.

The ability to communicate computer generated product definition data effectively between dissimilar systems is a pressing need in today's CAD/CAM community. This need is becoming even more critical as industry becomes more dependent on the computer in the factory of the future.

In order to get a better feeling for the problem, let us define what we mean by product definition data and present a brief scenario of how the need for a data exchange mechanism increases as the Engineering/Manufacturing world becomes more automated.

Product definition data is the data that is required and sufficient to describe a product to be manufactured. In addition to the basic shape (geometric) data, administrative (e.g., process planning) data, design analysis data, physical characteristic data, material processing data, and presentational data (e.g., general notes) are required (see Figure 1). Very often the data for engineering design consists of a subset of the product definition data, and manufacturing is required to supplement the engineering data.

Figures 2 through 4 describe a typical sequence of events as industries, in attempting to increase productivity and profitability, become more automated by increasing their CAD/CAM involvement. The need to communicate with internal systems becomes and continues to be a serious roadblock to increased productivity and cost savings.

In this presentation, we will address the area of the exchange of product definition data: what has been done, what is being done and what appear to be the directions for the future.

Before the advent of CAD/CAM Systems, product design information was communicated to manufacturing by the classical blueprint or engineering drawing. The drawing together with specification documents was enough to allow the product to be manufactured. Many of the "turn-key" systems were designed to produce engineering drawings to be passed on to manufacturing. Today, the blueprint is still widely used to communicate with manufacturing.

In our own "shop", not all design information is generated on our CAD/CAM Systems. Often a preliminary blueprint is generated from the CAD model and then completed on the draftsman's board before being passed on to manufacturing.

Many of the turnkey CAD/CAM Systems have been enhanced to include advanced features such as finite element modules (FEM) for analysis, numerical control (NC) milling models for generating cutter location (CL) data and solids modelers. McAuto's UNIGRAPHICS Product Line has introduced products such as GFEM, UNIAPT and UNISOLIDS to address these areas.

These new products and capabilities compound the data communication need, in the sense that not only do the systems (in-house or between contractors and subcontractors) differ, the versions of these systems often differ.

3. APPROACHES

What is the solution to this problem? How do we address the need to exchange data?

Basically there are four options:

1. Continue to live with the problem,
2. Agree to a single common data representation to be used by *all*,
3. Develop two-way translators for every pair of systems,
4. Develop two-way translators between each system and a neutral representation.

The first option is clearly economically disastrous and represents a severe block to technological advance.

The second option is probably the ideal, but totally impractical for proprietary and economic reasons. There are too many companies with too much money invested in too many systems.

It is instructive to compare the last two options. Given N systems, the third option requires the development of 2N(N-1) translators, the fourth requires 2N translators. On the surface, considering the "big picture", it seems clear that the fourth option is better. (See Figures 6 and 7)

However . . . this is still not the ideal solution. To see why not, consider what happens when translation is made between two systems.

Generally, in such a translation, the systems will have some representational overlap and some disjointedness. Entities not common to the two systems cannot be translated. To be sure, there are degrees of what may be called "not common", but the problem exists nevertheless. (It is in fact this problem which led to the opinion that Option Two is probably the ideal.) So, if any translation leads to lost information, what is the difference between Option Three and Option Four in this respect? To get from system A to system B in Option Four, one must translate from system A to the neutral system (possibly losing information) and from the neutral system to system B (possibly losing more information). (See Figure 8)

This effect of compounded information loss is unavoidable, or is it? What if the "neutral system" includes all the representations of all N systems? That would certainly avoid the problem of extra lost information. The maximum possible information would be transferred . . . but at the maximum possible cost in terms of translator development and processing efficiency. In fact, that situation is little different from Option Three!

The answer? There is no perfect solution.

So what is one to do?

4. HISTORY

Let's start by looking back at what has been done. The American National Standards Institute (ANSI) recognized in the early 70's that a problem existed. The formulation of a standard neutral format was discussed, and in June 1970 the ANSI Y14.26 Subcommittee was formed under the chairmanship of Ron Rau of McDonnell Douglas Corporation to address the Computer Aided Preparation of Product Definition Data. (See Figures 9 and 10) Dr. Shizuo Hori, also of McDonnell Douglas, became the chairman of Task Group Y14.26.1 in 1973 with the objective of generating a standard for the "Digital Representation of Physical Object Shapes". The scope of this task group extended beyond the geometry required for engineering drawings to that required to represent three dimensional physical objects.

The work of ANSI committees is primarily done on a voluntary basis and as a consequence usually extends over longer periods of time. (Figures 11-14 trace the activities of ANSI Y14.26 and some of the people involved.) During the period through 1978, the work of Y14.26 continued with additional task groups being formed to address both short and long term needs. One of those task groups was Y14.26.11 with the objective of addressing the standardization of digitized non-geometric product definition data.

The first draft of the Y14.26.1 task groups's work was published in 1976. Guided by ANSI editorial review board comments and assisted by a CAM-I grant to McAuto to improve the document, late in 1978, a proposed standard based on the work of Y14.26.1 entered the ANSI review cycle.

By August 1979, after resolving comments from reviewers, it was decided that following manual testing of the specification, it would be released as a draft standard for a one year public review period.

In this same time frame (September-October 1979), a second concentrated effort developed. Driven by the need to formulate an acceptable standard for use in communicating drawing and geometry data between commercially available CAD/CAM systems, funding was made available from the Department of Defense. The funding was administered through the Air Force Integrated Computer-Aided Manufacturing (ICAM) Program for the creation of an Initial Graphics Exchange Specification (IGES).

The key issue was to create an exchange specification in a very short time span. Boeing released its CIIN (CAD/CAM Integrated Information Network) Specification and General Electric released its Neutral Data Base to the public domain to be used as bases for IGES. Within three months, the three person task group from the National Bureau of Standards (NBS), Boeing, and General Electric had put together the first draft of IGES. In addition, and perhaps most important to the success of IGES, a committee framework was set up to provide technical advice on the development and use of IGES and to provide guidance, coordination and publicity within the CAD/CAM community.

Joint efforts of the IGES and ANSI Y14.26 committees during the Spring of 1980 brought together the work of the two groups into a proposed ANSI Standard. The geometric structures of the Y14.26.1 task group were embedded in the IGES file structure and included in the proposed standard as an alternative form for geometry. Figure 16 displays the contents of the Standard which was accepted in September 1981.

Even before the standard was approved, numerous CAD/CAM vendors had begun to implement Version 1.0 of IGES (i.e., Sections 2-4 of the Standard). In fact, IGES was a "defacto" standard. After the approval of ANSI Y14.26M in September 1981, many more vendors and users began to develop IGES translators.

On the other hand, to the best of our knowledge only McDonnell Douglas has implemented translators for Section 5 of the ANSI Standards (denoted ANSI5). Two way translators between ANSI5 and MDC's CADD (Computer Aided Design Drafting) System have been developed and tested. ANSI5 will be discussed later in the presentation.

5. IGES

When one refers to implementations of the Standard, one almost always means implementations of Sections 2-4 of the Standard (i.e., IGES). Figure 17 provides a simple description of how IGES processors work.

As early as Autofact 2 in November 1980, three vendors (Applicon, Calma, and Computervision) demonstrated initial IGES translator capability. At Autofact 3 in November 1981, Control Data Corporation brought the list to four. Three additional vendors, Gerber, McAuto-UNIGRAPHICS and Manufacturing and Consulting Services (MCS) brought the total to seven at MCGA in June 1982. Each of these demonstrations involved points, lines, circular arcs, conics and general notes. (See Figure 18)

Autofact 4 in November 1982 added IBM CADAM to the list of demonstrators and the complexity of the part had been increased to include linear, ordinate, angular and radial dimensions and labels. (See Figure 19)

These demonstrations have publicly exhibited an ability to transfer a subset of wireframe geometry and drafting entities.

The Autofact 5 demonstration part will be considerably more complex and will be dimensioned and displayed in four views. Twelve vendors have expressed their intention to participate in the demonstration. (See Figure 20)

According to the August 1983 IGES Newsletter: "To date, a total of thirty-one vendors have announced plans to provide IGES Implementation". Figures 21-23 give a partial list of vendor implementations compiled in Autumn 1982.

What does all of this say about the ability to exchange product definition data between dissimilar systems? Do simple demonstrations such as these indicate that the exchange problem has been solved? Not really!

IGES is not perfect. No one ever said it was. Sections 2-4 of the Standard have known deficiencies. IGES has been criticized for its rather inefficient file structure. The data depicted by IGES is drawing data rather than complete product definition data. There are major application areas (e.g., printed wiring boards, finite element modeling, piping, manufacturing technology and solids) that are not addressed in the Standard. There are some unclear, ambiguous and misleading definitions in Sections 2-4 of the Standard document. Sections 2-4 of the Standard don't handle rational parametric curves and the associated surfaces.

It must be pointed out that, although we have been referring to Sections 2-4 of the Standard as IGES, IGES has really gone well beyond the Standard.

The committee framework that was established early in 1980 has gone on and remained extremely active. As a volunteer effort composed of representatives from both the vendor and user community, the committees have met on an average of four times a year. Their principal objectives have been to repair errors, oversights and ambiguities in IGES, to extend its coverage of CAD/CAM application areas, and to provide means of testing and evaluating IGES processors. See Figures 24-29, which describe the objectives of the two principal IGES committees – Extensions and Repairs and Test, Evaluate and Support.

Version 2.0 of IGES was published in February 1983. In addition to corrections of typo's, omissions, and errors and clarifications of ambiguities, extensions were made in the areas of finite element modeling, electronic printed wiring boards and generalized surface representations. An optional binary file structure was also included in Version 2.0. (Figure 30)

Areas still under consideration by the IGES committees include extending into the area of solids modeling, manufacturing technology, and plant design, enhancing the FEM and electrical extensions, and building in the ability to reference library files. (Figure 31)

Some concerns of committee members are the issues of upward compatibility of later versions of IGES, the modularization of IGES related to application areas, and the question of whather IGES should be regarded as simply an exchange specification or as a database structure requiring carefully defined back-pointers. These issues and others have impacted the development of IGES.

IGES has indeed become a "household" word. Many government contracts, especially in the aerospace industries, are specifically calling out IGES compatibility as a requirement in their statements of work. Consequently many users are demanding that translators be available for their CAD/CAM Systems.

What is meant by IGES compatibility is still a rather ill defined concept. For example, is it valid to claim that a vendor is IGES compatible when his translator handles only a subset of IGES (e.g., wireframe geometry)? The level to which vendors have successfully implemented IGES is quite varied. Efforts are underway to establish testing procedures for validating IGES translators. One of the objectives of the Test, Evaluate and Support Committee is to develop a test library and test procedures. In addition, the Product Definition Data Interface Project about which we will talk shortly is addressing this need.

6. ALTERNATIVES

Can IGES solve both the near term and long term need to exchange product definition data? Is it the best available exchange specification? Is it the only exchange specification available in the public domain?

The first two questions are certainly open for debate. The third question can be answered definitely no.

Vought Corporation which deals in aerospace structures and assemblies developed the Standard Product Data Exchange File Format (SDF) which predates IGES. Vought describes their format as:

1. System and application independent
2. A complete, accurate representation
3. Compact
4. Independently extendable
5. Minimizes data translation
6. Minimizes associativity pointers

It has a packed binary file structure with sequential data records.

SDF is regarded as being sufficient for all of Vought's current applications. They are committed to continuing to develop interfaces from their applications to SDF and to expanding SDF as future applications. (See Figures 33-34)

Interfaces between SDF and CADAM, PATRAN and CADDS-3 currently exist and are used at Vought.

The Standard Interchange Format (SIF) was developed by Intergraph in 1979, primarily to meet the needs of users in the areas of mapping, petrophysical exploration and facilities management. Characteristics of SIF are: (See Figure 35)

1. Free format
2. Both ASCII and Binary forms available
3. Inclusion of attribute information as well as graphics information
4. Key word driven
5. Easily edited
6. No pointers

Intergraph, too, is committed to its exchange specification. All external formats are converted to SIF before entering their system.

Translators between SIF and Applicon, Autotrol and Calma exist and are used.

It should be noted that both Intergraph and Vought are developing IGES translators as well. At Intergraph, they have found it to be more efficient to convert their IGES files to SIF.

Both Vought and Intergraph have done studies which indicate their formats to be more efficient than IGES.

A third alternative to IGES as a neutral format is Section 5 of the ANSI Y14.26M Standard. Section 5 addresses geometry only and has the following characteristics: (Figure 36)

1. Geometry represented as relational structures
2. Rigorous definitions
3. Topology tied to underlying geometry
4. Minimal number of structures
5. Extendable
6. Curves, Surfaces, and Solids represented

Once again the efficiency of this approach is noted as greater than that of IGES geometrical representation.

Criticisms of Section 5 have usually centered on the fact that non geometry is not considered and that the underlying mathematics of Section 5 is too difficult to understand and implement.

The first criticism is addressed by referencing an approach to non-geometry in an ANSI5-like structure that was presented by H. H. Shu of McDonnell Douglas Astronautics Company (MDAC) at the NASA Symposium on Geometric Modeling at Langley, Virginia, in April 1983. Although not implemented, the approach was straightforward and in the spirit of ANSI5.

In answer to the second criticism, McDonnell Douglas has successfully implemented ANSI5 translators.

Comparative results using CADD/IGES and CADD/ANSI5 translation on a variety of parts from simple to rather complex once again pointed out the inefficiency of the IGES file structure.

IGES has emerged as the most prominent candidate to fill the need for a neutral exchange specification. It is being implemented by virtually all of the major vendors and has drawn much interest in the international CAD/CAM community. (See Figures 37-38).

Nevertheless, it still primarily addresses blueprint data, rather than the complete geometric model and complete product definition, and validation procedures are not available for benchmark verification of IGES.

7. PDDI

The Air Force has maintained their interest in the importance of product definition data exchange, especially as it relates to the aerospace industry.

In order to accelerate the development of an exchange specification which would address the communication of complete product definition data between Engineering and Manufacturing, the Air Force ICAM office, on October 1, 1982, awarded McDonnell Aircraft Company the Product Definition Data Interface (PDDI) Project.

The PDDI Project consists of two tasks as illustrated in Figures 37-42. Task I considers the communication of CAD drawing data from one graphics system to another by means of translators based on Sections 2 through 4 of the ANSI Y14.26M Standard (IGES Version 1.0).

Task II considers the specification of "complete" part definition as input to manufacturing and the communication of this data to both engineering and manufacturing functions using today's blueprint, including process planning, N/C programming, quality assurance, tool design, etc. The entire program, started in October 1982, will span a 33 month period with Task I running concurrently with Task II during the first 12 months.

Task I consists of the development of validation procedures to verify the implementability of IGES and test procedures to determine the current level of implementation of IGES translators. A principal part of Task I is the administration of these test procedures at numerous vendor/user sites and the analysis of the results which will be input to Task II. The development and administration of Task I procedures is being handled by Booz, Allen & Hamilton, Inc. – a principal subcontractor on this project.

A principal by-product of Task I of the PDDI Project is the development of a unified set of test procedures for the validation of IGES translators. During the course of the testing, twelve vendor sites have been visited and their IGES translators have been tested. Test parts, much more complex than those used at AUTOFACT and NCGA, were used. The test parts depicted typical aerospace parts of the machined, turned, sheet metal, and composite type. The test results are being compiled in a generic fashion to be used as input to Task II of the project. The procedures and IGES test files will be released later to the public domain to be used by CAD/CAM users and vendors for validating IGES translators.

Task II is the major thrust of the PDDI Project. Here long term manufacturing needs will be defined, a state-of-the-art analysis will take place, system requirements will be specified, and a prototype project definition data interface will be designed, developed and demonstrated.

The key focus in Task II is on the Engineering/Manufacturing Interface, especially as it relates to the aerospace industry. A primary goal is to eventually eliminate the blueprint and the need to convert engineering data to manufacturing form. (See Figure 45) This will significantly reduce Engineering/Manufacturing cycle time.

Driven by manufacturing needs data, a PDDI exchange format will be developed. Some examples are the needs for part control information, detail part geometry, detail part features, dimensioning and tolerancing information, material requirements, finish and processing specifications, and drawing notes. (See Figures 46 and 47) An example of the improved design/manufacturing productivity resulting might be illustrated in Figure 46 where the drafting terminology of lines, circles and curves are interpreted functionally as loads, flanges, pockets, fasteners, and holes.

Figure 50 describes the PDDI deliverables:

1. A *PDDI Exchange Standard Format* for communication of Product Definition Data,
2. A *Working Form* of the PDDI format to be used in the prototype demonstration,
3. *Access Software* to allow external systems to extract appropriate data. (For the prototype demonstration, this access software is targeted to be simple in form.)

Figure 51 displays the PDDI Project Organization. It should be noted that MCAIR has teamed both manufacturing and engineering personnel to provide a key blend of expertise. The MCAIR team is completed by (1) Booz, Allen & Hamilton, contracted to develop and administer the test procedures for Task I assisted by the Task I Working Group consisting of key leaders from the IGES Committees, (2) United Technologies contracted to lend expertise on turned parts, (3) the Task II Industry Review Team, chaired by J. F. Lardner (John Deere) and consisting of Industry Leaders to comment and advise the MCAIR Team, and (4) McDonnell Douglas Automation Company to lend system design and implementation expertise.

The PDDI project has the potential to significantly affect the CAD/CAM community by providing a needed interface for the communication of "complete" product definition data which will serve as input to the factory of the future. (See Figure 52)

8. ROADBLOCKS

The PDDI project has set lofty goals and directions in addressing some of the needs for the aerospace "factory of the future". However, many variables enter into the solution of the data exchange problem.

Product Definition Data

- Administrative
- Design/Analysis
- Basic Shape
- Augmenting Physical Characteristics
- Material Processing
- Presentational

McAuto

GP31-1728-24

FIGURE 1

The Need for a Data Exchange Mechanism

Users - Typical History

Action	Results
• Procurement of Interactive Graphics Drafting System	• Increased Productivity Over Manual Methods
• Addition of High Speed Plotter, Same or Different Supplier	• Increased Productivity • Minor Compatibility Problems
• Further Procurement of Drafting Systems — Same Supplier - New Generation — Different Supplier	• Extended Capability • Additional Compatibility Complications

McAuto

GP31-1728-45

FIGURE 2

The Need for a Data Exchange Mechanism

Users - Typical History

Action	Results
• Interface to User's Internal Database and/or Applications Programs	• Extended Utility • Further Compatibility Problems
• Other Divisions Procure Same or Different Systems	• Extended Application • Further Compatibility Problems
• Write System to System Translators	• More Efficient Communication of Data • N(N-1) Possibilities

McAuto

GP31-1728-46

FIGURE 3

The Need for a Data Exchange Mechanism

Users - Typical History

Action	Results
• Develop Standard Data Formats for Internal Data Communications	• Reduce Proliferation of Translators • Expensive to Develop • Expensive to Maintain • Inefficient • Increase in Productivity
• Interface With External Systems — Supplier — Customer — Other Divisions — Etc	• Compatibility Problems • Interface Possibilities Are Infinite

McAuto

GP31-1728-47

FIGURE 4

Exchange Options

- Maintain the Status Quo
- Universal Data Representation
- Two-Way Direct Translators
- Two-Way Translators to Neutral Format

McAuto

GP31-1728-33

FIGURE 5

Direct Translators (2n(n – 1))

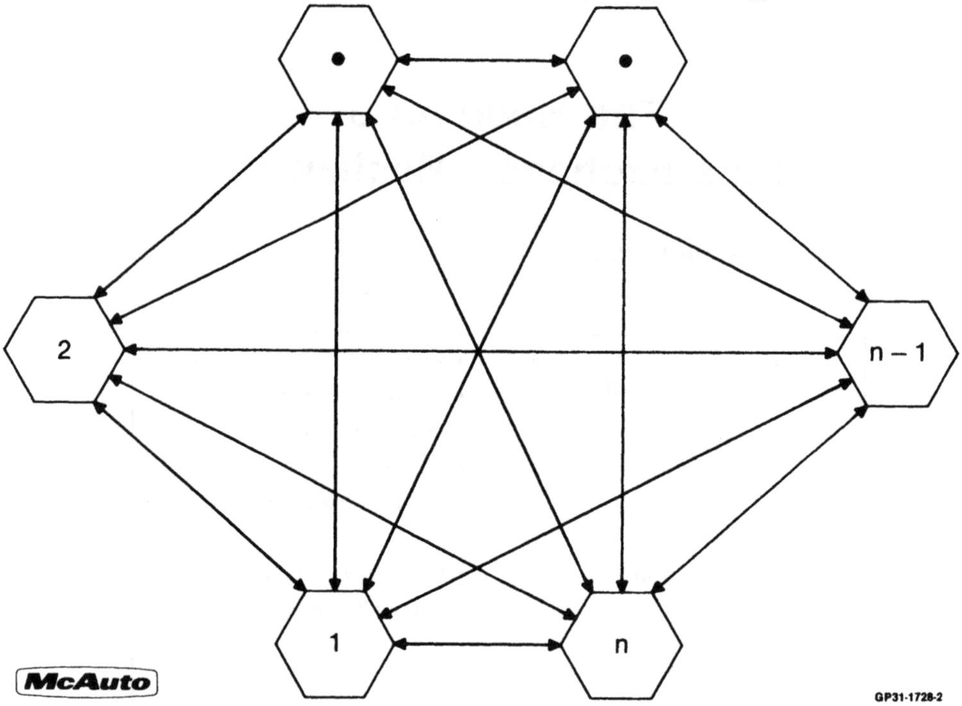

McAuto

GP31-1728-2

FIGURE 6

Neutral File Translators (2n)

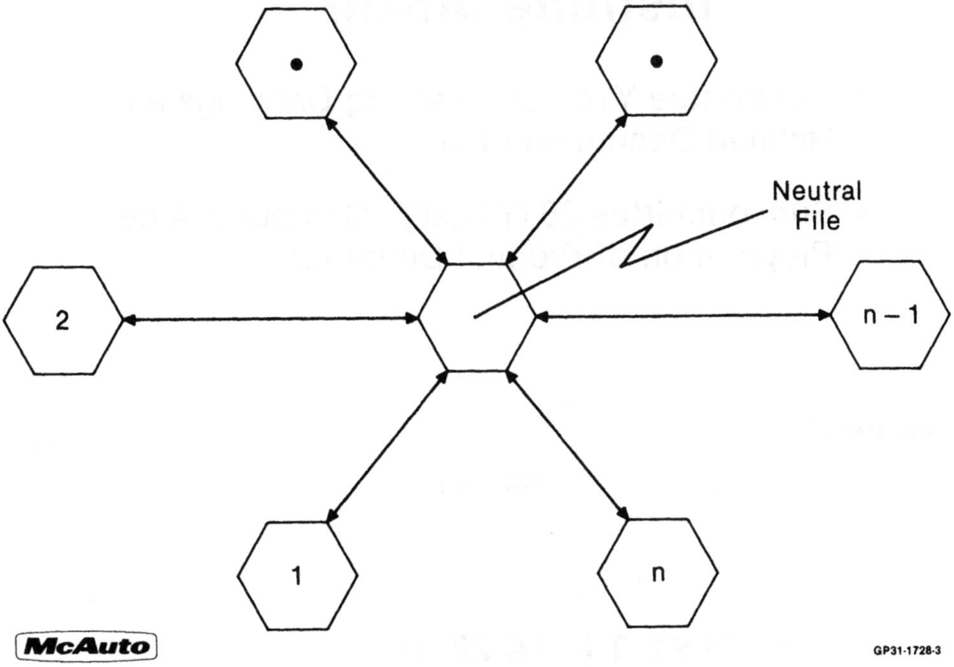

FIGURE 7

Information Loss in Translation

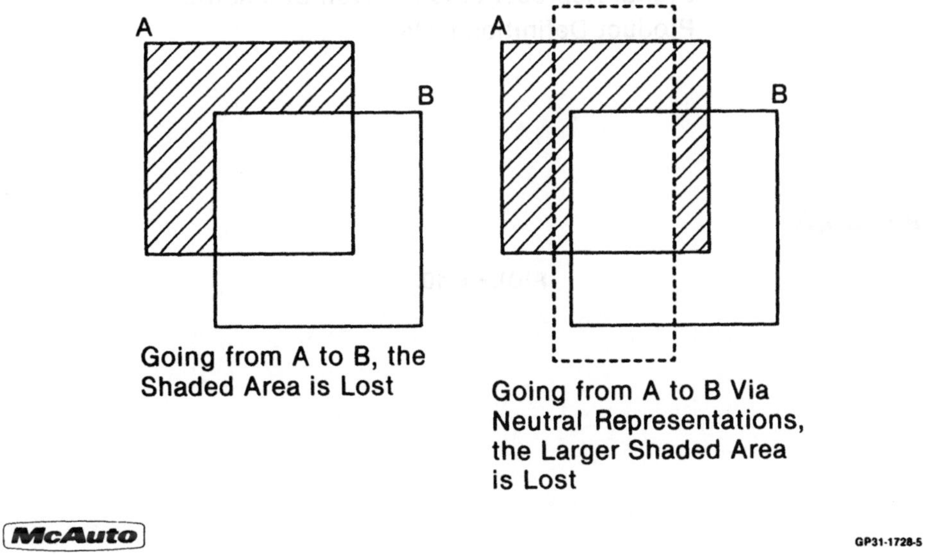

FIGURE 8

American National Standards Institute (ANSI)

- Committee Y14 - Engineering Drawings and Related Documentation
- Subcommittee 26 (Y14.26) - Computer Aided Preparation of Product Definition

McAuto

GP31-1728-34

FIGURE 9

- **Task 1 (Y14.26.1)**
  - — Digital Representation of Physical Object Shapes
- **Task 11 (Y14.26.11)**
  - — Digital Representation of Non-Geometric Product Definition Data

McAuto

GP31-1728-35

FIGURE 10

ANSI Y14.26/Task Y14.26.1 History

- June 1970 - Established R. W. Rau, Chairman
- January 1973 - S. Hori Chairmanship Task 1
- July 1973 - Changed to 3-D Approach

McAuto

GP31-1728-32

FIGURE 11

ANSI Y14.26/Task Y14.26.1 History

- August 1976 - "Digital Representation of Physical Object Shapes" Published
- April - June 1979 - ANSI Public Review
- January 1980 - IGES/Task 1 Draft Standard Agreed to Study Merge With IGES

McAuto

GP31-1728-23

FIGURE 12

ANSI 5 - Its Predecessors and People

- January 1973 - Chairmanship of Task Y14.26.1 to Dr. S. Hori
- April - June 1979 - ANSI Public Review of Document
- November 1979 - CAM-I GMP Research by McAuto. Major Contributors: Dr. R. F. Emnett, Dr. H. H. Shu, E. A. Werner

McAuto

GP31-1728-31

FIGURE 13

ANSI 5 - Its Predecessors and People

- January 1980 - Revision of Y14.26.1 Based on GMP Research
- March - June 1980 - ANSI 5 Developed. Team: Dr. M. F. Collins, Dr. R. F. Emnett, Dr. R. L. Magedson, Dr. H. H. Shu
- November 1980 - March 1981 - Drafts Edited and Comments Resolved (Technical Team, Dr. S. Hori, and E. G. Houghton)

McAuto

GP31-1728-22

FIGURE 14

IGES History

- **September 1979 - Department of Defense Establishes Need for Standard**
- **October 1979 - IGES Committee Established**
  - Boeing CIIN to Be Starting Point
- **January 1980 - Version 1.0 Released**
- **May 1980 - Joint Standard (IGES/ANSI 5) Proposed to NBS**
- **September 1981 - Standard Accepted**
- **February 1983 - Version 2.0 Released**

McAuto

GP31-1728-21

FIGURE 15

ANSI Y14.26M Standard

Digital Representation for Communication of Product Definition Data

Section	Title
1	Introduction
2	File Structure (IGES Version 1.0)
3	Geometry (IGES Version 1.0)
4	Non Geometry (IGES Version 1.0)
5	Geometry (ANSI Y14.26.1)
6	Non Geometry (ANSI Y14.26.11) (to Be Added Later)

McAuto

GP31-1728-36

FIGURE 16

IGES Environment

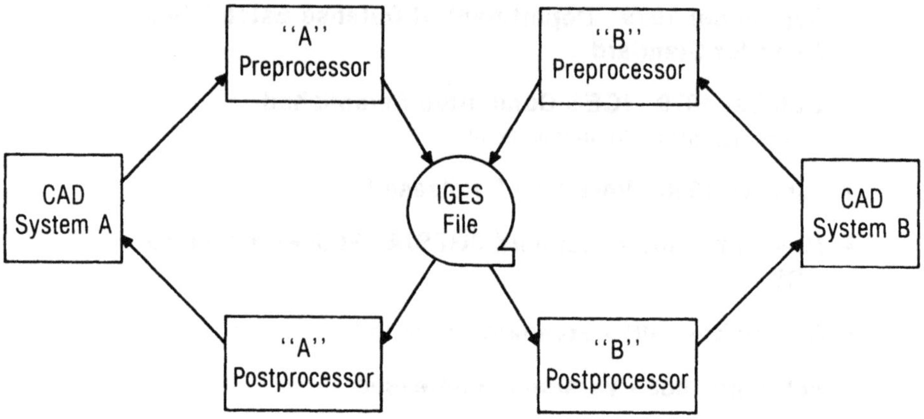

McAuto

GP31-1728-4

FIGURE 17

Sample Part

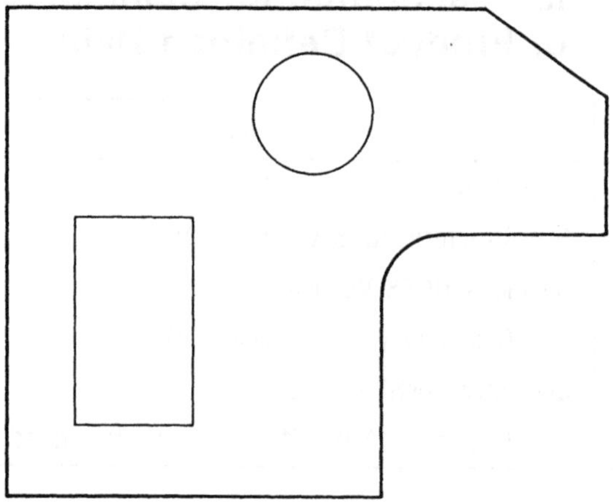

McAuto

GP31-1728-37

FIGURE 18

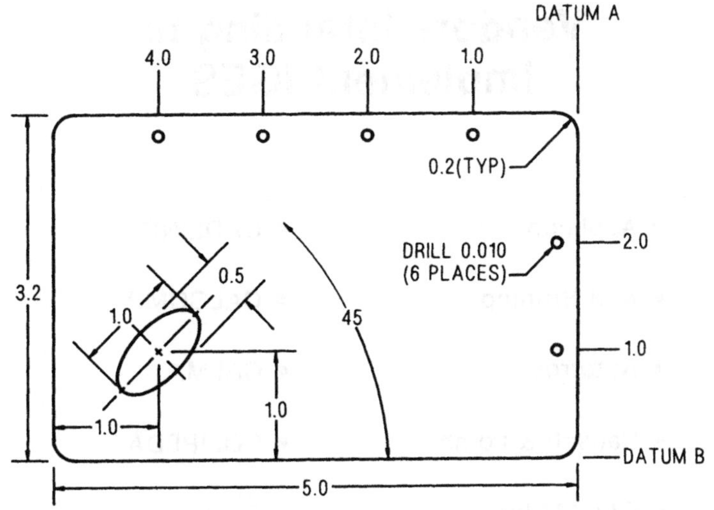

McAuto GP31-1726-38

FIGURE 19

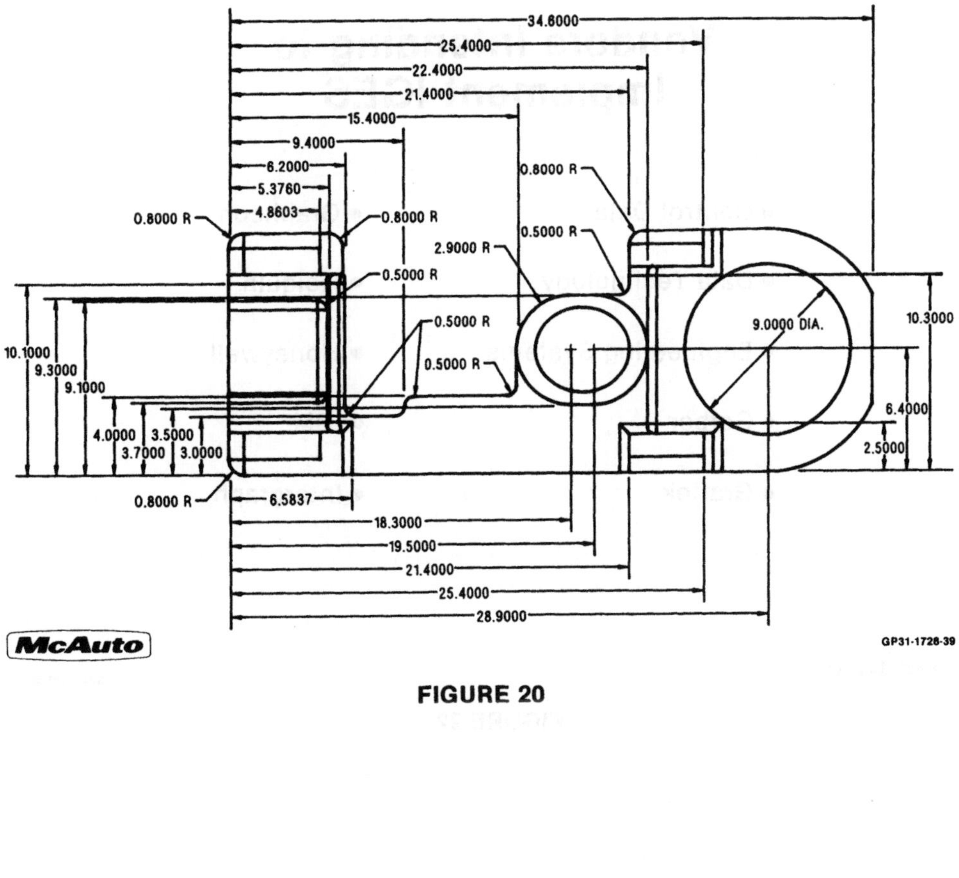

McAuto GP31-1726-39

FIGURE 20

Vendors Intending to Implement IGES

- Applicon
- A M Bruning
- Autotrol
- Bausch & Lomb
- CADAM Inc
- CADLINC
- CALCOMP
- CALMA
- COMPEDA
- Computervision

McAuto

GP31-1728-46

FIGURE 21

Vendors Intending to Implement IGES

- Control Data
- Data Technology
- Engineering Systems
- Gerber
- Graftek
- Graphcon
- Holquin
- Honeywell
- IBM
- Intergraph

McAuto

GP31-1728-53

FIGURE 22

Vendors Intending to Implement IGES

- K & E
- McAuto UNIGRAPHICS
- MCS I
- NCS
- Prime
- Project Software
- Summagraphics
- Systemhouse
- Tektronix
- T & W Systems

FIGURE 23

GP31-1728-54

IGES Committee Structure

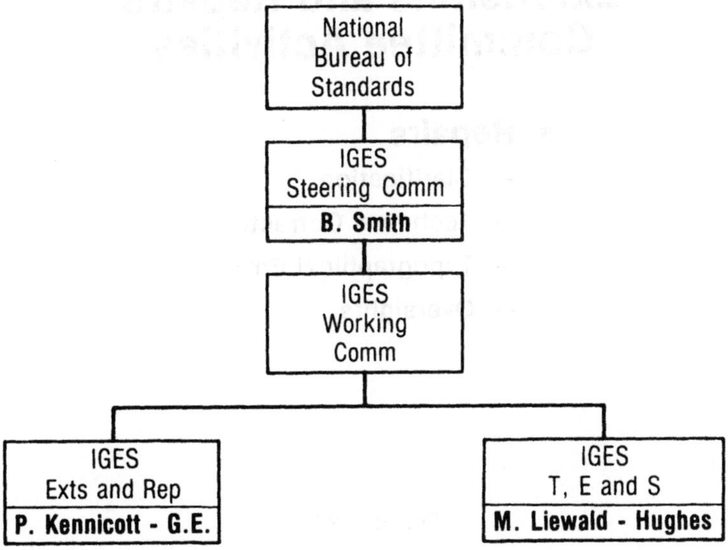

GP31-1728-1

McAuto

FIGURE 24

IGES
Extensions and Repairs Committee Activities

- **Extensions**
  - Application Areas
  - Generalization
  - Efficiency Enhancements

McAuto

GP31-1728-19

FIGURE 25

IGES
Extensions and Repairs Committee Activities

- **Repairs**
  - Clarification
  - Technical Corrections
  - Typographical Errors
  - Oversights

McAuto

GP31-1728-20

FIGURE 26

IGES
Extensions and Repairs Subcommittees

- Advanced Geometry - W. Gruttke, McAuto
- Electrical Extensions - D. James, Hughes
- Drafting - R. Schmid, McAuto
- Finite Element Modeling - R. Ivey, Westinghouse

McAuto

GP31-1728-30

FIGURE 27

IGES
Extensions and Repairs Subcommittee

- **Manufacturing Technology - R. Lord, MCAIR**
- **Plant Design - P. Rourke**
- **Miscellaneous Ad Hoc Committees**
  - — Binary File Structure
  - — Reference File Library
  - — Text

McAuto

GP31-1728-18

FIGURE 28

IGES
Test, Evaluate and Support Committee Activities

- **Test Library**
- **Recommended Practices Guide**
- **Glossary**
- **Analysis of Testing**
  - Autofact Demonstrations
  - User/Vendor Exchanges

McAuto

GP31-1728-17

FIGURE 29

IGES
Version 2.0 Additions

- Electrical Extensions
- Generalized Surface Extensions
- Finite Element Modeling Extensions
- Optional Binary File Structure
- Glossary

McAuto

GP31-1728-16

FIGURE 30

2

INTEGRATING ENGINEERING AND MANUFACTURING

William D. Beeby
William D. Beeby and Associates
Kent, Washington 98031

INTRODUCTION

The Boeing Commercial Airplane Company uses computers in every phase of airplane production. After a long gradual struggle, Boeing has managed to harness the efficiency and speed of the computer and to channel its abundant capacities for work to perform a mind-boggling variety of jobs.

In the past, communication among the varying disciplines of airplane manufacture was simply a matter of walking a few feet from one person to the other in the same building. Oral communication sufficed. Thanks to the advent of more and more specialization, the geographic separation of our people and activities, the sheer magnitude of our productivity, and the diverse languages and processes our computers use, we have had to devise new ways to pass along information to one another. With the development of an integrated network and a companion management system, Boeing's CAD/CAM community has been able to provide solutions to its communication problems (Figure 1).

CAD/CAM has probably seen the most widespread and spotlit progress of all the postwar industrial endeavors. Significantly, its use has forced an integration among systems and disciplines to an extent never experienced before. Trend watchers have enjoyed monitoring the transitions and innovations of batch CAD, part and template making, master dimension, and drafting using the interactive graphics computer. And now they are observing the unprecedented marriages of like and unlike systems, design and operations, data management and production functions, and the cooperative ventures of industry and university professors and students.

The communication of information at Boeing is chronological with airplane engineering and manufacturing processes. First, geometric data for the basic structural characteristics of the airplane are developed in preliminary design. In the engineering process, detail parts that are represented on drawings are developed. These drawings become the controlled source of information for the airplane design and manufacture. The network encompasses this information from beginning to end, and is, therefore, not merely a storage or communication device, but includes all of the modules of the system.

DEFINITION

CAD/CAM is a broad term covering the use of computer systems for design or manufacturing purposes and for communicating design or manufacturing information. All CAD/CAM applications use, in varying degrees, the three special computer functions of interaction, graphical visualization, and large-scale data communications. A little over a decade ago, CAD/CAM was the exclusive province of large aerospace and automotive companies. Now the situation has changed dramatically.

With the mushrooming of computer systems throughout the industrial domains, there arose a need for a coordinating function whose purpose would be to correlate requirements and uses of systems and to oversee the burgeoning computerization among disparate activities. In 1972, Boeing formed its CAD/CAM support organization. This new group, as in other companies, saw that computer program data bases were developed, that systems were consistently maintained, and that the personnel who would run them were trained.

CAD/CAM is universally recognized as holding the greatest potential for productivity improvements in manufacturing. Its effect is perhaps more staggering and more pervasive than the Industrial Revolution. A truly integrated CAD/CAM or computer integrated manufacturing (CIM) system provides computer assistance to all business functions from marketing to delivery. It plays a part in planning, engineering, operations, and management, reaching across organizational boundaries. A successful CIM factory provides a vital key to the development of methodologies for manufacturing and advances in productivity.

METHODOLOGY OF INTEGRATION

A full integrated CAD/CAM system involves the use of processes and of systems in such a manner that the output of one system serves as the input to another. For example, at the business planning and marketing level, a customer order organization receives input from the sales people describing the product. The product description serves as an output to the engineering design function. If the product contains previously designed components, a computer-assisted drafting system would output the engineering drawing information to the planning system. The product description, if it contains new components, would serve as input to a computer aided design system that uses interactive graphics as a design aid to produce engineering and manufacturing information. Complete implementation of CIM results in the automation of the information flow in a business organization, from the entry of an order through every step in the process.

HISTORY

Design aids in the form of analysis programs were among the first uses of digitial computers, and with the addition of output plotters they provided the ability to generate and visualize geometric information. The term "computer aided design" became recognized following the work at MIT in the late 1950's on multiaccess computing, and produced the computer-driven cathode ray tube display and the time sharing computer systems that made interactive computing possible at a reasonable cost. Time sharing led to a massive growth in the use of multiaccess computers for both interactive and remote batch processing in engineering, since it has made possible online working and interactive computing.

Structural design and parts programming for numerical control machining were explored in aerospace applications in the 1960's. The first computer aided design application involving graphics was in the design of printed circuit boards, eventually expanding to complex multilayer configurations. More recently, engineers are combining the graphical description of complex geometrics with analysis programs such as stress calculations and dynamic response.

The developments that have recently taken place in kinds of display terminals, plotters, communication equipment, auxiliary storage, and in minicomputers means that it is now possible to analyze an application requirement in detail, either from a specialized vendor system, or from inhouse development.

DEVELOPMENT OF CAD/CAM AT BOEING

When the B & W airplane was built in 1916, Boeing had a total of 21 employees. The whole operation took place in the two-story frame building that Boeing called the "Red Barn." The engineering department was upstairs; the production shop was downstairs. The machinist could talk directly to the engineer about his design. Communications were excellent and the drawing became an almost after-the-fact record of their agreement. At that time, the technique worked. Technologies advanced over the next decades when Boeing produced the 747 and other commercial and military jet aircraft, and now, the 767 and 757 airplanes. The Red Barn became a manufacturing complex spread throughout the Puget Sound area with seven million square feet of shop space.

Boeing began using computers to aid in the building of airplanes in 1958. At that time, manufacturing created APT programs to describe parts shown on engineering drawings. That data could then be used to create a punched paper or mylar tape and make machine parts. This procedure is well known as numerical control or NC machining.

Computing has come a long way since 1958

and the digital computer and the aerospace industries are, in fact, contemporaries and Boeing has played an important part in the development of both disciplines. The result of this dual technology is total reevaluation of relationships: quality control to methods, facility to remote facility, system to unlike system, veteran to new hire, and manufacturing to engineering. In the last few years, we have experienced an upsurge of new communications between and among fuctions and peoples who have barely spoken to one another in the past, whose responsibilities rarely touched. We at Boeing have been bridging gaps as never before—and it is this integration among disparate factors that I would like to address and share with you today, for it is in this new communion of functions that we will find the keys to the future.

COMMUNICATIONS AND CONTROL NETWORK

To date, no one has implemented a completely integrated CAD/CAM system. It requires time and a certain amount of risk-taking to build up total integration. At Boeing, we've combined several systems that have proven successful in their own right to begin our integration effort. Some, such as the batch CAD program, have been around for decades; others are fairly new and we've tied them in to their connecting processes from the beginning. Actually, the most difficult and time-consuming task has been to convince the organizations to sit down and admit that their operation is going to run a lot better and more cost effectively—which, in today's environment means survival—if they look upstream or downstream and join forces with the other operations.

The CAD/CAM network enables design data to be communicated between the various processors from design to manufacturing. Design data are currently generated in three ways: master dimensions data from the TX95 MD system on the CDC Cyber; APT input from the batch CAD systems in the IBM 3033; and interactive graphics devices. Each of the various systems in the network may have its own method of representing geometric entities, its own language, and its own data format.

Computer programs called preprocessors and postprocessors have been developed that will convert goemetric and dimensional data as they are are passed between the various systems in the network via the Geometric Data Base (GDB) (Figure 2).

In July 1976, batch CAD data were transferred directly to an interactive system via the CAD/CAM Integrated Information Network. One of the objectives of this netowrk is to allow manufacturing to use the computerized geometry, developed by engineering projects, as input to batch programs. Following are some of the kinds of systems that must communicate:

- o Large-host IBM
- o Cyber systems
- o Various standalone minicomputers
- o Batch programs and interactive computer graphics
- o Graphics and NC machines.

Due to the geographic distances of the systems that generate and utilize design data, a communication network is required. The interactive graphics systems can communicate with the CDC Cyber and the GDB. A direct link exists between master dimensions and the geometric data base as both are located on the CDC Cyber, which can also communicate with the IBM 3033. The IBM 3033 system is used by both engineering and manufacturing for APT processing. Engineering design data orignating as batch CAD or APT can be communicated back to the Cyber and ICG systems.

When the CAD/CAM Integrated Information Network was conceived, there was no computer-based data management system that could properly manage, store, and allow retrieval of large variable-length files of data. The network system was set up to include transfer of geometric data through a plotting subnetwork, data base archiving, diagram change notice support, graphics interface improvements, data management systems, and global network extensions to include Wichita, Vertol, subcontractors, partners, vendors and clients (Figure 3). One of the most important improvements in the management of data bases is accessibility by other affiliated disciplines such as finance, sales and marketing, purchasing, and customer support, which do not routinely perform design and manufacturing work.

Engineering information developed at remote locations must eventually be communicated to the central data center. From there, the information is transmitted to the other locations where it is required.

For example, engineering information created at an interactive graphics work station is transmitted to manufacturing where numerical control creates the part itself. It should be emphasized that remote systems have different equipment with different communication and data format requirements.

CIIN is capable of communicating and sorting drawings representing installations, assemblies, and detail parts including the geometry and annotation describing it, as well as the geometric definition, plotting data, and source data. These geometric and drafting data are stored in the Geometric Data Base Management System (GDBMS) in three different formats: a standard format, a native format, or a format that supports APT partitioned data sets (PDS).

SYSTEMS MANAGEMENT

To automate the functions involved in engineering design and manufacturing process planning, a general purpose parts description method suitable for all design, planning, and control requirements is essential. The method must be capable of describing part geometry and form in three dimensions through a high-level language. The part models must be stored in a form that permits rapid retrieval and display, and the part data must be structured so that the association of parts to assemblies, and to manufacturing processes, can be simply and rapidly made.

The roles of personnel involved in engineering design, production planning, manufacturing engineering, and associated functions are changing. Computer integrated manufacturing is putting more manufacturing know-how in the hands of engineers as they design and develop products. As a result, the boundary between engineering and manufacturing is beginning to fade. CIM will require an increase in skill levels. Instead of performing repetitious tasks, workers will be doing setup, changeover, and maintenance operations. Managers will have to develop the skills needed to effectively manage and motivate a more highly skilled work force.

THE CAD/CAM MODULES

The computer integrated factory comprises modular subsystems, each controlled by computers that communicate with one another and form a distributed computer system. The intelligence from each of the setups, though they vary in language and functions, will be translated and understood by a central large-host computer, the communication link.

PRELIMINARY DESIGN

Taking geometric information from the Geometry Control System (GCS), designers begin the final design. The preliminary design GCS loft provides the basis for the Master Dimensions systems and the airplane's final design. The GCS creates such data as the kind of airfoil required to make the plane fly, basic lift and drag capabilities, wing size, etc. (Figure 4).

If the engineer wants to look at, for example, temperature versus altitude on a certain pressure probe in the engine, those two variables can be selected and within ten seconds the system displays and plots the information. The system uses a PDP 11-70 computer and refresh graphics. Using initial information from the technology staffs, an engineer working with an interactive terminal can feed preliminary information such as wing plan form and thickness, engine type, and so forth, to define an initial airplane configuration geometry. The computer performs size-matching calculations. The following are examples of other preliminary design tasks that the computer can do:

- o Kinematics of trailing edge flaps
- o Testing noise levels in decibels
- o Simulations
- o Temperature variations

Graphs drawn on the system can print from the scope with the quality required for the FAA or for handbooks, so that these kinds of data do not have to be drawn by hand. For example, in a graph that shows input to study separation characteristics of the space shuttle, you could actually see it move because you have the aerodynamic characteristics of the two units. By applying both of them, you see how they would react.

GEOMETRY CONTROL SYSTEM (GCS)

The Geometry Control System (GCS), a computer program system that numerically defines airplane external geometry, provides preliminary configuration data for wind tunnel testing. Numerous configurations are modeled in wind tunnel tests. The wind tunnel group uses computers in two distinct applications.

MODEL DESIGN

The model design group uses an interactive computer graphics system to design the model airplanes for testing in the transonic and other wind tunnels. This geometric design information is used in drawing form to make the 14-inch to 20-foot long models that are manufactured on the site. The interactive design system allows rapid creation of drawings resulting in faster turnout of models, an important feature because wind tunnels are expensive and it is not cost effective to have them sitting empty.

PROJECT DESIGN

When loft surface requirements come back from the wind tunnel and preliminary design, loftsmen mathematically create a surface definition. They begin writing batch CAD or TX95 extractions to cut contours. It is the CAD group's responsibility to create software packages for jobs the airplane projects want them to do. Requirements from marketing, preliminary design, tech staffs, wind tunnel, project design are all still being analyzed. One group builds a design, another tears it down. Input from all groups must be considered. All elements must be woven into a coherent pattern.

Beginning from layouts, project designs toward the mockup release. For the first time they create detail designs. They settle on a final centerline drawing, then decide stringer spacings. In addition to the Master Dimensions definitions (MDDs), conversational time sharing (CTS), and other design programs, all serve as design aids in mathematically defining the components of the airplane. The general and special batch CAD programs described in the section Assembly and Installation Design contain a tremendous amount of geometry, which manufacturing is able to use for tooling and fabrication of parts. These programs are located in the large-host mainframes.

Part cards and drawing sheets are indications of the amount of design work and geometric entities. Historical data from the YC-14 are available for analysis and comparison. These data have been accumulated in four major areas of endeavor: Structures, Payloads, Systems, and Propulsion. Developmental work is done with various kinds of interactive systems. Each of these systems is successful in specific applications. This conclusion was reached in part by studies and partly through trial and error. We are gaining from our experiences in a production environment.

ASSEMBLY AND INSTALLATION DESIGN (FIGURE 5)

PARAMETRIC DESIGN

The batch process, using the APT language, was originally a manufacturing-oriented technique that translates NC programmers' instructions into machine tool motion commands, which is still its major function. In recent years, the cutter has been substituted with a pen and APT has been used in computer aided design operations as well.

Batch CAD is a parametric design process combined with the Master Dimension lofting programs. In this mode, some parts are stored on the computer so that manufacturing engineering can access the structural geometry of the aircraft parts. Batch programming is very effective for contour-related families of parts, such as inspar ribs, body frames, leading edge ribs, and so forth. (Figure 6). During the development of the YC-14, batch programs were initiated that pioneered the use of batch CAD. However, due to geometry differences and software construction, these programs were not readily usable on subsequent aircraft design programs. Following this experience, the central engineering organization at Boeing began creating several large general programs that incorporated macros (parts or components that are preprogramed). The general programs could then be adapted to any Boeing commercial aircraft configuration. These first general batch programs were for wing box structures and body monocoque.

Special batch programs -- those that were created and not taken from the stockpile of previous general programs -- were written for the fixed leading edge, the leading edge slats, the fixed leading edge machined ribs, and the trailing edge flaps. These are ideal special batch applications because they are families of look-alike, contour-related parts. The basic geometry for either the actuators or the tracks can then be combined with the contours for a perfect geometric fit. A program for the whole series of leading edge ribs and slats can be created with the contour as the only basic variable (Figure 7).

Input to APT, formatted in a formal APT source language, defines the geometric entities that make up an aircraft component and also the machine tool motion commands that machine the part itself. Five sections of the APT program govern its execution: control, translation, arithmetic, editor, and postprocessor sections. APT source files comprise two separate kind of data: geometry definition statements and the tool motion commands. The advantages of APT programming are its rapid production of many diagrams that vary slightly from one another. This benefit was realized recently most clearly in the creation of the wing box on the 767 with its 26 ribs. Using APT programs, we were able to create and save for future use whole families of parts.

INTERACTIVE CONVERSATIONAL SYSTEMS

To solve a section analysis problem for, say, a body frame, a Boeing design engineer has merely to sit down at a computer terminal (Figure 8). Instead of using a slide rule and a calculator, the designer can use either a portable desk top terminal or an alphanumeric scope terminal. These calculations are available on the large-host IBM 3033 computer. About 1973, Boeing developed some conversational analysis programs that allowed direct access to call up conversational design aid programs and general batch and special batch programs. Before that time the only computing support for designers was large batch programming. These conversational programs eliminate a lot of back-and-forth checking and rechecking and long waits, delays that were demoralizing to engineers and devastating to design schedules. As a result, designers often returned to manual methods.

Conversational programs save engineers as many as 100 hours for each hour spent defining section properties of complex shapes. Simple mechanism kinematics take ten times as long to calculate manually as they do using the conversational systems. This kind of productivity is accomplished with programs that write out questions in designer language -- that is, English, just as one designer talks to another. To make it easy for the user, the program asks only three kinds of questions: those that expect a yes or no answer, a multiple choice number, or numeric values such as dimensions, location and magnitude of loads, and coordinates of points. Even a person with minimum typing skills can use these programs effectively. No computer programming skill is required. When the designer is satisfied that the design meets the requirements of structural integrity of the airplane, then it goes to the structures technology staff.

Conversational programs perform the following analysis:

- o two-dimensional linkage kinematics
- o fatigue and load transfer of structural joints
- o extrusion and selection
- o mass properties such as weight and center of gravity
- o landing gear performance
- o rib and spar web analysis
- o second degree curve generation and display
- o bearing and lug design and analysis
- o spring design and analysis
- o General purpose plotting and digitizing

INTERACTIVE COMPUTER GRAPHICS

In 1974, the Boeing Commercial Airplane Company brought in the first interactive computer graphics systems. The advent of interactive graphics to Boeing marked the beginning of a new epoch in computer-aided design because of its reciprocating abilities. This dimension allows designers to "talk back and forth" -- to interact -- with the computer and to realize the results of their work immediately. The use of interactive graphics has magnified the scope of the use of computers, and the effectiveness of batch programming has narrowed the gap between the design and manufacturing communities and has opened up new possibilities for excellence in all design projects, such as structures, systems, payloads, and propulsions.

Each of the interactive system installations provides an autonomous computer system that enables a user to tap the systems's interactive capabilities to prepare engineering drawings, to store them either on a disk drive or on magnetic tape, and to display and revise the drawings.

Geometric entities include points, lines, circles, conics, strings, splines, tabulated cylinders, surfaces of resolution, view matrices, and groups.

Some entities include crosshatching, centerline, linear dimensions, angular dimensions, and so forth. Other capabilities include a variety of entities that support NC machine tool control operations, finite element modeling, layering and leveling, view selection, polar coordintes, point spacing control, size, scale, and protection codes. Another ICG function is annotation of drawings.

DRAFTING

We now have a data base of general and special batch programs on file whose worth is immeasurable. Using a combination of batch programs and interactive graphics, drawings such as skin panels and stringers can be manipulated to be presented on standard engineering drawing forms. These large parts (some 65 feet long) would be unmanageable to detail entirely in a batch mode to produce full-scale drawings.

The process of drafting on the computer saves us an immeasurable amount of effort and paper, not to mention storage space. On the wiring diagram program, for example, the same drawing goes out to a dozen different customers, but only the effectivity block has to be changed. And by joining the batch drawings to an interactive system, we can take any number of identical drawings built on parametric design, and label them on the graphics computer. We can store one master design for the grahpics and text files on different layers or different hosts.

The drafting process itself takes place on the flatbed plotter, the computer output microfilm (COM) unit, or on the electrostatic printer for a check print.

AUTOMATED DIAGRAM SYSTEM (ADS)

Project groups use an ICG system to draft and revise wiring diagrams, schematics, and Automated Diagram System staffs have created large libraries of parts such as electrical symbols, light modules, and instrument panel indicators. They have perfected time-saving execute files to perform unwieldy and lengthy procedures. Printed circuit board design and drafting for both 767 and 757 are atuomated.

Wire bundle and wire connection properties contained on a digitized diagram are compared with similar data output by the Wire Information and Release System (WIRS).

Drawings are postprocessed and sent to a photocomposer where the drawings are recorded on 35mm microfilm.

MARRYING BATCH CAD AND INTERACTIVE GRAPHICS

Often the most efficient method is to combine the flexibility of interactive graphics with the efficiency of batch CAD as the former can be readily revised, and as the latter can create geometry faster than anything we have ever found.

Having written a batch program, 60 leading edges can be created overnight, for example. The next day we can have all 60 of them drawn. But the batch process does an inefficient job of putting words on the drawing. To make something unique on any one of those sixty drawings and to program it in batch mode is too unwieldy. But through marrying the two processes, graphics can do the unique job.

GROUP TECHNOLOGY

One of the hardest tasks in establishing an effective data base has nothing to do with computers, but rather with revising company procedures. For example, a weights engineer wants data structured to give the volume automatically, whereas a structures engineer needs yet another kind of data. To cope with these disparities in requirements, a method of part classification, called *group technology*, was devised. Group technology is a philosophy in which similar parts are grouped together into part families to take advantage of their similarities in design and manufacturing. Group technology recognizes that similar decisions are repeated throughout the design and manufacture of a product and encompasses a systematic classification and grouping of subsets of the products.

Group technology is a process whereby production is viewed in terms of logical families of parts that can be produced within a single line or group of machines and people, rather than routing a workpiece from the milling department to the drilling department to the grinding department, and so forth. There are two main advantages in using group technology: workpieces go into and out of a line or cell in one to two weeks rather than wandering around a plant from department to department for three to four weeks before they are completed. This means quicker scheduling and better control of the workpieces at all times. The other advantage comes in knowing who is responsible for quality of the finished workpiece.

The difficulty in adopting a group technology system lies in grouping the product into families of parts, which usually means arriving at a logical coding system. It means altering procedures and methods that have been used for decades and reorienting personnel to a new production control concept.

It is practical and economical for the first time to manage complex product design and manufacturing processes by taking into consideration the total production system. The cost of implementing systems of group technology is not accrued in the computing software. By far the greater cost is associated with the development of the classification and decision logic that uniquely describes each company's product and process. No major benefits can be realized from these techniques without the dedication of significant resources to evaluate the current method of operation, determine optimum or preferred solutions, and construct logical, hierarchical statements of those decisions. (Figures 9 & 10)

DETAIL DESIGN
(Figure 11)

An example of an effective use of the computer in a detail design process is Tubend, a process for bending tubes. We now have numerically controlled bending equipment that can bend hydraulic tubing to fit the hard points–blocks, fittings, and clamps. Tubend is one of the first computer aided design and manufacturing systems at Boeing that represents an entire system. Integration is the primary difference between Tubend and most other CAD systems. Using Tubend, the designer can route a hydraulics system on a roadmap, a diagram showing how the components are interconnected. The Tubend system, a tutorial one that asks questions of the user that will allow input of design information, aids in stress, thermal, performance, troubleshooting, and other kinds of analysis. Tolerance buildup is eliminated. Numerically controlling bending equipment works directly with the geometric tube coordinates released from Tubend. Tubend loads the geometry into data storage in the production shop and checking is done using Tubend-generated mylar templates.

WIRE INFORMATION AND RELEASE SYSTEM (WIRS)

WIRS is an online network using CRT terminals in engineering design and operations for the design, manufacture, assembly, and testing of airplane wire bundles. WIRS provides configuration accountability through the engineering mockup and planning phases; support for assembly and test of wire bundles; automatic generation of shop paper; reduced data errors and flow time; and more economical fabrication and installation of wiring.

WIRS makes it possible to manufacture some 700 wire bundles per day composed of 32,000 wire segments and to process 500 wiring changes to accommodate customer-unique features. Engineering data and bundle design data are entered into the WIRS data base, which is used by manufacturing to build up the bundles. WIRS also feeds parts ordering data to other systems that order the needed parts. In the wire shop, this system provides instructions of data for numerically controlled machines to cut and label each wire, and in some cases, lay the wire to form a bundle. Other bundles are laid up manually according to printed instructions from the computer. All bundles are tested by computer, using test plans generated by WIRS and hookup instructions are produced to support installation by shop personnel.

The final step is the extensive documentation of the airplane's electrical system needed by the customer and WIRS supports this requirement as well. The system gives us the configuration control and accountability of wiring that we had lost, with reduced flowtime and increased accuracy at all stages of the design-to-installation cycle.

WORK BREAKDOWN STRUCTURE

Another effective system that is a spinoff of automation is called the *work breakdown structure* (WBS). WBS is a systematic and orderly arrangement of elements of activity. It is a numbering system that allows anyone working with a product or assembly–from preliminary design to final assembly—to easily and quickly identify and locate a part. This ingenious and simple device operates on a system of levels, the first level being an alpha program designation for an entire assembly. The second level designates the product's major functional elements, such as structures, systems, propulsion, payloads, and so forth, using an airplane as an example. The third level further designates major assemblies, such as wing, body sections, environmental control and avionics systems.

DESIGN RELEASE AND PLANNING (Figure 12)

Boeing taps several related online data base technologies to increase productivity of manufacturing engineers and production control and factory personnel. These include the online planning system (OLP) and the parts inventory control system (PICS). In 1980, an interface between OLP and PICS was implemented to provide a mechanical load of blueline and lot time orders for the model 767 airplane and soon after, the 757. PICS has been expanded to support the aerospace and military programs and provides such data as locations of parts, accountability, and availability of parts.

At Boeing, the OLP system maintains and updates manufacturing plans for 2,000,000 parts and tools and releases 7,000 orders each day to the shop in accordance with these plans and needs. We have almost 900 terminals that input or access data in the OLP system. This system gave us a 25% to 30% improvement in productivity by our planners.

Sitting at a terminal and using a few simple commands and a menu, the planner has access to process plans that list the material required, components necessary for an assembly or tool, and a narrative description of the step-by-step process of making the part. Instructions are given for flame cutting, rough cutting, and anodizing, for example. These plans are divided into two formats, for shop and for planning. They vary only slightly. Shop formats include a flow time report and the planning format lists mnemonics and revisions. Both shop and office locations house the IBM terminals and printers necessary to use the system.

The parts inventory and control system (PICS) is another online system that tracks parts availability for almost one million details and assemblies built within Boeing. Through 1.2 million transactions each month and a variety of reports, PICS is invaluable in maintaining on-time production schedules.

With the use of such systems as the foregoing, part shortages fell from a level of more than 4,000 each month to fewer than a hundred.

TOOL DESIGN AND NUMERICAL CONTROL (Figure 13)

In the shop, the APT language translates part definitions into drawing instructions for NC machine tools. Preparation of APT input is accomplished by the numerical control distributed system (NCDS), which enables production engineers using terminals in the part programming areas to interact with a minicomputer.

NCDS provides for job editing and remote job entry (RJE) to the IBM 3033 for APT processing. Output from APT is returned to NCDS. In some cases, telecommunication lines connect NCDS directly to machines for transmittal of machine control data. This is called a direct numerical control (DNC) connection.

GRAPHICS AUTOMATED TEMPLATE SYSTEM (GATS)

The GATS process was the first integrated system that crossed the engineering and manufacturing boundaries with the same data. Engineering creates a flat pattern, and either sends the information over a communication link to Auburn and the template maker there creates part geometry on the graphics system, or a photo contact master (PCM), which is a full-sized copy of the engineering geometry, for the shop to use. Then, using a number of computer aids, such as a menu with libraries of often-used commands in tooling tabs, identification tabs, scribe lines, bend lines, and cutout information, the template maker computerizes the template. Once the template geometry is defined, the system automatically generates an NC tape for the NC mill. A four-by-eight foot sheet of template stock is fastened to the bed of the machine and the template is cut using routers, drills, and a scribing tool. All of the equipment, from the graphics work station to the NC machine, is together right in the shop. (Figure 14)

Since an interactive graphics system was introduced into the template shop operation, flow time and direct labor hours have been reduced by 75 percent for template manufacture. Digitizing, if required, data set translation, cutter path generation, numerical control tape generation, and plotting take about one-half hour. The Burgmaster takes about 20 minutes to machine a template. A 6- to 12-hour manual job now takes fewer than 45 minutes. (Figure 15)

Many routines and processes are using computer graphics, but the one that holds the most promise, in terms of time and effort saved, is the template-making function. GATS is CAD/CAM at its finest. The job that had once taken several days can now be done in half an hour.

The process of producing templates may sound complicated, but with the use of GATS and the Burgmaster, it takes mere minutes and is accurate to a tolerance of ±.005. It is a shop-controlled computer graphics system located in the same building where the template is actually made.

Initially the GATS equipment was operated in the tool design area, away from the shop where its end product would be used. But now all of the equipment from the graphics work stations to the NC machines are together right in the shop and important human engineering factors—lighting, efficiency, and comfort of the operator—have been addressed.

GATS is not the only integrated system that the fabrication group is developing. The optical comparator template system makes use of computer graphics and a photo plotter to produce an inspection template for use on a device for projecting, magnifying, and checking small, often dime-sized, parts that the micrometer cannot reach. Parts that use CCTs are gears, splines, and timing mechanisms for hydraulics and plugs, for example.

The NC tube bending and Yoder roll programs use the computer to provide data for bending ducts and forming sheet metal. The universal weld table program performs a variety of layout and calculating functions for designing duct weld jigs. The mill cutter design program is used to design helical cylindrical mill cutters. These cutters are eventually used for machine stringers and spars.

Tool designers use the mill fixture design program to design tools that locate and hold such engineering parts as one-of-a-kind machine castings and forgings. To aid the designer, a standard library of mill fixture components and other special routines, such as PEP and execute files, has been created. PEP—parametric element processing—is a language similar to both APT and FORTRAN.

The duct weld jig design and machining program uses a combination of interactive graphics and APT to generate duct weld jig designs including the NC tape to drill the tool base.

DETAIL FABRICATION
(Figure 16)

GRAPHICS AUTOMATED FLOOR PANEL SYSTEMS

Another gap to be bridged is between design and shop. Some of the same data pass between engineers and shop personnel. An example of an unbroken flow of information between design and shop is the process of making floor panels for the Boeing line of commercial aircraft. The first step in the life of a newly created floor panel data set takes place on the design side, with the creation of the data set. A CAD/CAM release unit sends the data set over the CDC Cyber to the Auburn Fabrication Division. Next, the fabrication planning group generates an assembly plan and tool order for the shop via the Online Planning System, which tracks down manufacturing information.

Next, graphics operator extracts a copy of the engineering floor panel data set from the geometric data base, then processes the data set over using the GAFS routines. The operator produces the APT source, which is sent to the IBM large host computer for post-processing. The cutting information generated is loaded onto the Onsrud library for use by direct numerical control.

DNC sends the NC information via microwave to the Barracuda router in the production shop and the panel is machined; its nesting, that is, layout placement, information is stored in the library for possible later use. (Figure 17)

Before drilling, drill-pattern preparation is made without templates. The data generated for locating the fastener holes in the floor beams of the aircraft are used to guide a router that produces matching holes in the floor panels thereby ensuring a perfect fit when the parts are brought together in final assembly.

At Auburn, nested floor panels are automatically drilled on a flatbed vacuum-chuck router under numerical control. Here, the nesting arrangement determines the drilling order selected by the computer program.

It all comes together in final assembly where automatic floor drilling equipment, using servomotors controlled by a microprocessor, locates and drills fastener holes. This automated process uses engineering and manufacturing information to guide the drilling equipment through the structural frame.

SUBCONTRACTING
(Figure 18)

Using the advanced technologies of electronically passed data over long distances, companies are becoming more able to take advantage of other companies' expertise and equipment in subcontracting work to them. Examples in industry of such cooperative ventures abound. On the 757 Program, for example, Boeing called on such companies as Avco, Rockwell, LTV, and Fairchild to take Boeing's geometric data and to use them as baselines for designing tools. In this instance, it was a relatively simple matter to send these data by magnetic tape. (Figure 19)

Improvements to the subcontracting process in the future will come with more effective procedures to allow a more complete understanding of the data and how they are to be used. Creating workable procedures and mutual understanding constitute greater problems to be solved than the technological capabilities do. To be effective on an airplane program, it was necessary for CAD/CAM contributions to reach outside Boeing. About 50 percent of the 757 airplain, including major body structural sections, was built by other firms. (Figures 20 & 21) While the CAD/CAM capabilities of the subcontracting firms varied, we found that it was efficient for them to use Boeing's computer data for the geometry of the parts they built. Boeing sent more than 2,000 data sets to the subcontractors. In one instance, the subcontractors were able to have their computers talk directly to Boeing to extract detailed geometric data. Such an approach saved flow time and unnecessary tooling costs.

SUBASSEMBLY
(Figure 22)

AUTOMATED FLOOR DRILLING

The main deck of the 747 contains 345 panels and 23,000 fasteners that require 28½ man-hours per aircraft to install. Drilling 23,000 holes in the floor is a highly repetitious process that makes it a made-to-order computer application. Engineering locates the fastener holes to comply with customer requirements. Working from the engineering design, planning, tool design, NC programming and tool estimating and scheduling organizations, design metal-drilling templates for each of the 345 panels.

An automated floor drilling procedure was established when engineering began transmitting their geometric data to the large host computer to make them available to the shop.

A drill unit was designed that takes instructions from a microcomputer. Servomotors move the drill heads on the drill unit framework, also directed by the microcomputer. This automated system drills 30 holes per minute for a cost savings of 85% over the manual drilling operation. (Figure 23)

ASSEMBLY AND INSTALLATION
(Figure 24)

The assembly of wing spars, which consist of an upper and lower chord, webs, and stiffeners, has been greatly improved with the use of CAD/CAM. Such assembly requires drilling and installing 3,000 fasteners, a process which, if done manually takes almost 100 templates to locate the holes. Using equipment that operates on CAD/CAM data, the fasteners are drilled and installed working from engineering geometric data. This approach is cleaner, more accurate, and faster than conventional methods. (Figure 25)

The riveting process for wing panels has also been automated. This equipment locates and drills holes, countersinks, inserts, and shaves some 30,000 fasteners from the 757 wing. CAD/CAM data sets have enhanced this process to the extent that the first 757 wing was completed without a single location error.

Production of mockups on our two newest programs reached a high level of quality and we've seen that same level in the first production airplanes as well. The wing-stub join on an earlier program—the 747—took two weeks. On the 757, the wing-stub join was completed in 18 hours. Boeing beat manhour estimates for assembly of the first 767 airplane by 30% to 60%. Eventually, mockups will probably be eliminated entirely, so that this kind of improvement in precision and quality is vital.

IMPACT ON THE ORGANIZATION

The very nature of the manufacturing organization is to resist change. The functional groups: design, administrative control groups, customer support, sales and marketing, research and development, tool engineering, QC, operations, planning, and plant engineering become easily polarized.

Some companies attempt to manage their systems using steering committees instead of establishing administrative procedures and altering the organizational structure to better align functions that will be affected by the systems. Often, failure of a steering committee occurs when a committee is used as a substitute for badly needed managing, administration, overall structuring, and comprehensive strategies.

The first step CAD/CAM planners must take is to ask how systems should be organized. Responsible managers must decide how many mainframe centers there should be and to whom they report. They must also determine how development groups should be organized, how many there should be, and where they should be located. The following questions should also be asked:

1. Should CAD/CAM be pulled out as a separate organization?
2. How should CAD/CAM relate to other computing systems such as office and information systems?
3. Should the lines of reporting for operating groups be altered in direct reporting or "dotted-line" relationships?

These questions cannot be answered until someone sorts out the responsibility for CAD/CAM resources of every participant. Each one must have a clear understanding of goals, objectives, and responsibilities.

When getting started, in each area of activity it is better to begin with the simpler applications. Prospective users of CAD/CAM must first understand where the most effort occurs. Progress to more complex operations can come about as understanding of needs, successes, and the practical applications of systems increases.

SELLING THE COMPANY ON CAD/CAM

The following are some tried-and-true methods when presenting a CAD/CAM proposal to the many individuals and representatives of the varying functions CAD/CAM will touch.

- o Free Samples: One selling technique used effectively to sell people on using any new idea is to offer a free sample. A new project frequently operates on a very tight budget and if you tell people involved that you'll do something for them for free, they can't resist.
- o Slush Fund: A requirement to begin a program is a cooperative and informed management who will carry a side budget, a kind of slush fund. Include training of personnel in that slush fund.
- o Object Lesson: Carry around a tangible part to illustrate the accuracy possible with a data set generated by engineering and used by manufacturing.
- o Status Reporting: Always and frequently keep management informed.

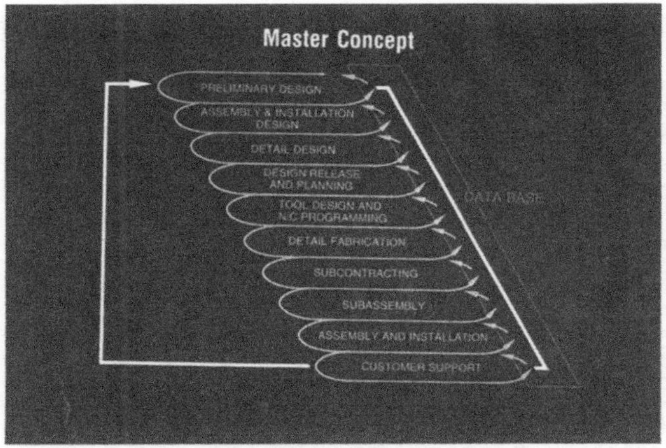

Figure 1

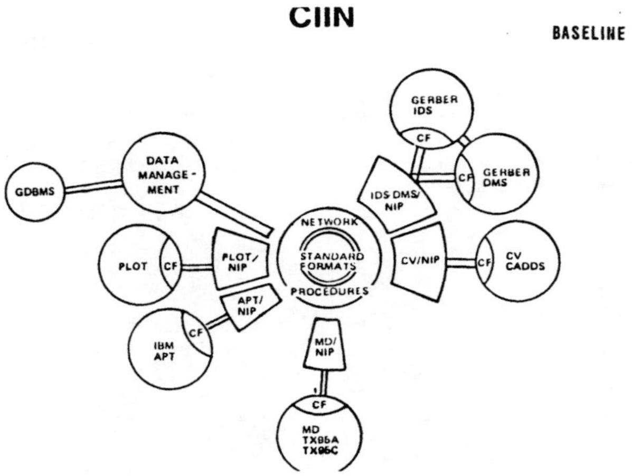

Figure 2

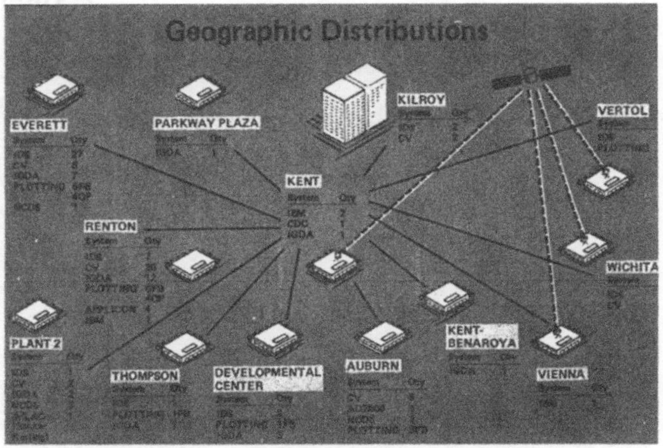

Figure 3

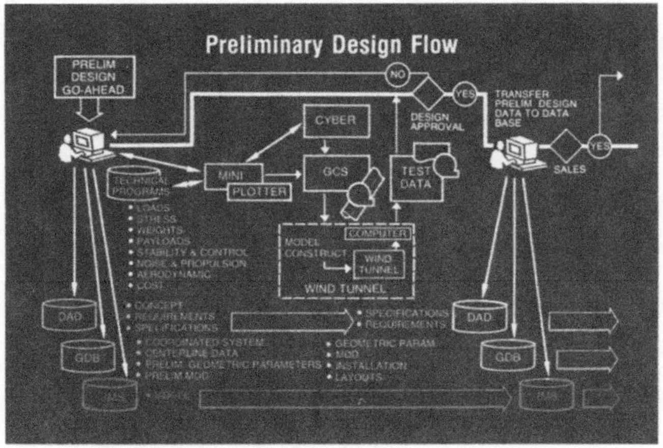

Figure 4

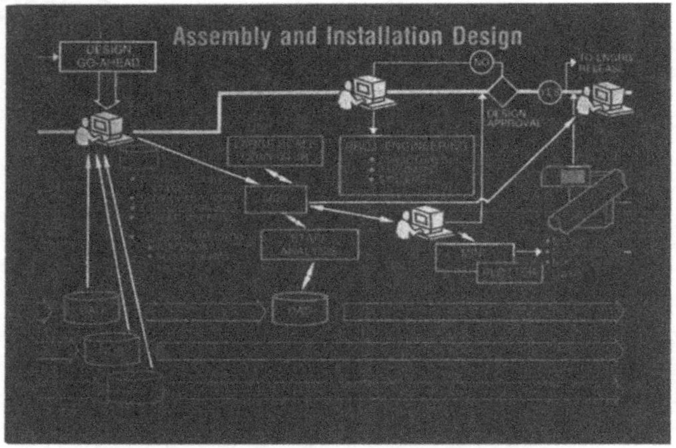

Figure 5

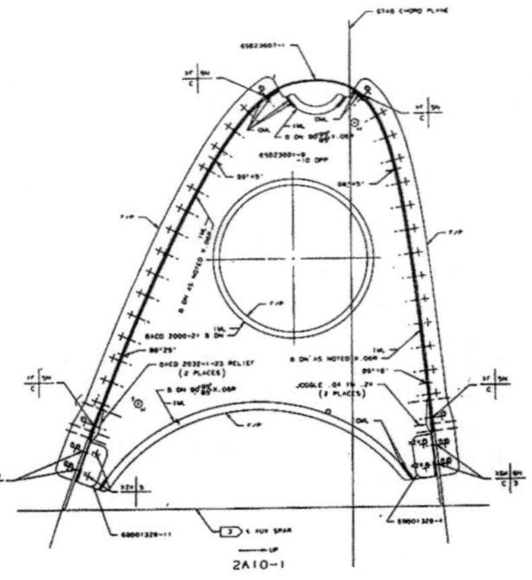

Figure 6

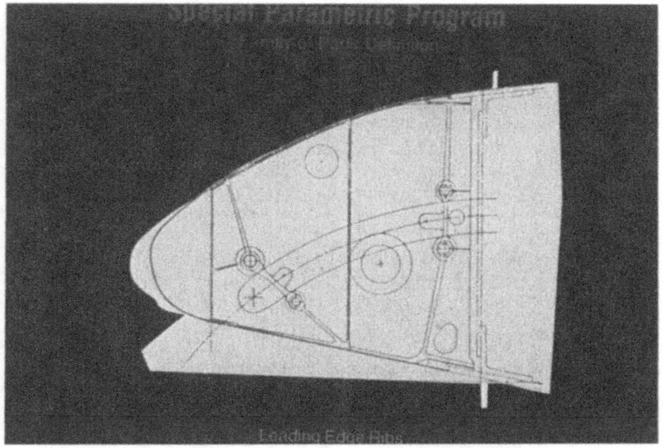

Figure 7

Figure 8

BENEFITS FROM GROUP TECHNOLOGY

- IN ENGINEERING
  - ACCESS PRODUCIBILITY INFORMATION
  - REUSE EXISTING DESIGNS
  - ACCESS DESIGN DATA
  - IMPROVED MATERIALS AVAILABILITY DATA
  - ACCESS MATERIALS PROPERTIES
  - SUPPORT DESIGN STANDARDIZATION
  - REDUCE ADCN/DCN COSTS
  - DESIGN TO COST
  - DESIGN TO WEIGHT

Figure 9

BENEFITS FROM GROUP TECHNOLOGY

- IN MANUFACTURING
  - STANDARDIZATION OF PROCESS PLANNING
  - INCREASE IN LOT SIZE THROUGH PARTS GROUPING
  - STANDARDIZATION OF RAW MATERIAL SIZES
  - REDUCTION IN RAW MATERIAL WASTE
  - REDUCTION IN SET-UP TIMES
  - PROVISION OF MEANS OF UTILIZING FAMILY TOOLING
  - REDUCTION IN WORK-IN-PROCESS
  - REDUCTION IN PROCESS FLOW TIME
  - OPTIMIZE EQUIPMENT PROCUREMENT AND UTILIZATION
  - FACILITATE COST AND PERFORMANCE ANALYSIS
  - IMPROVED WORK CONTENT ESTIMATION

Figure 10

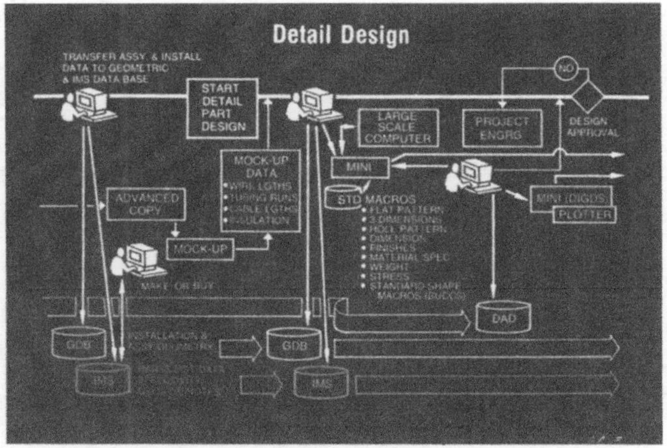

Figure 11

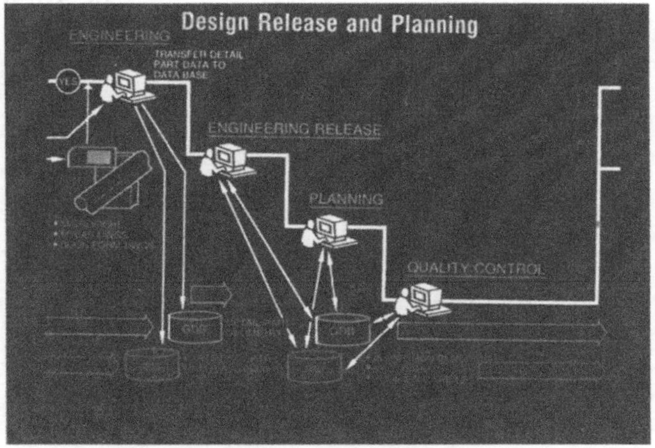

Figure 12

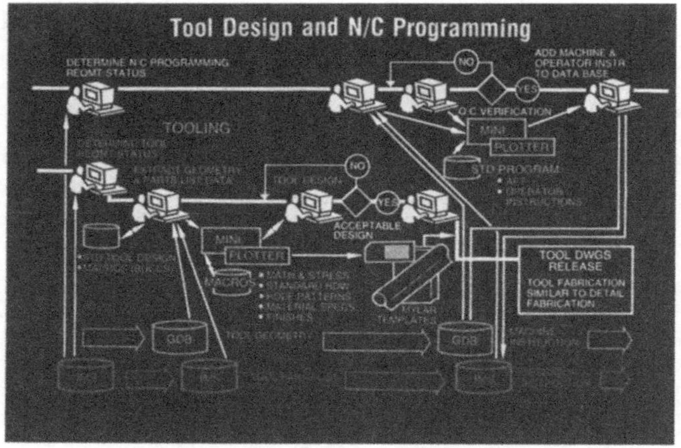

Figure 13

Figure 14

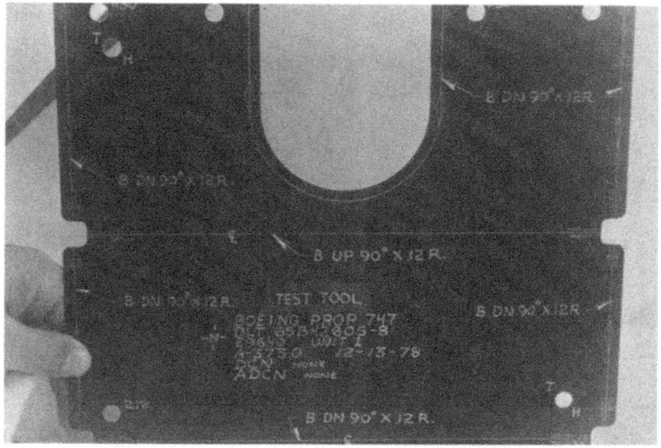

Figure 15

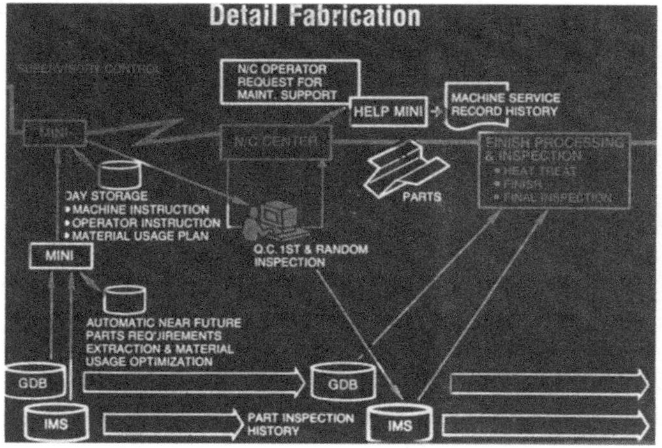

Figure 16

Figure 17

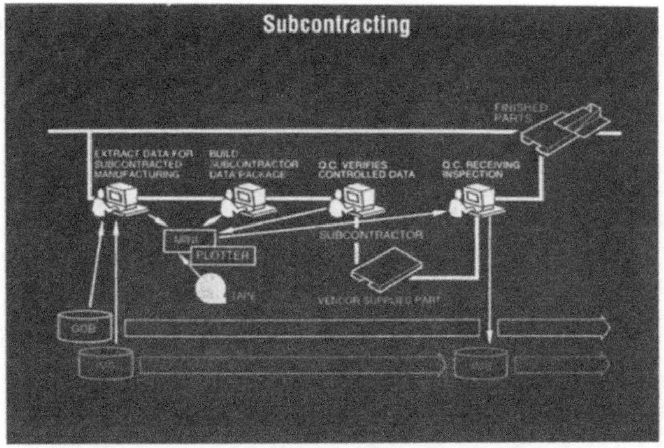

Figure 18

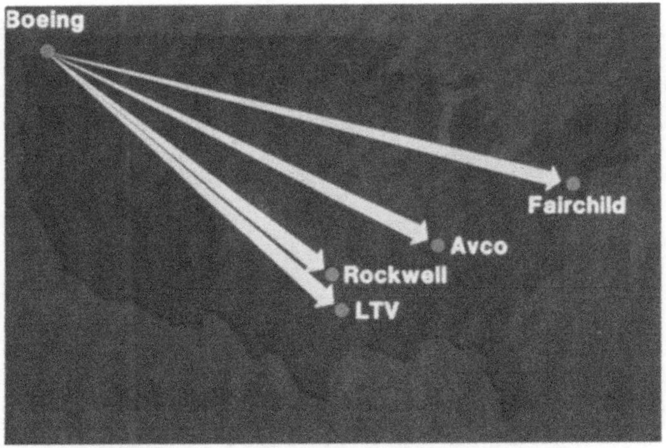

Figure 19

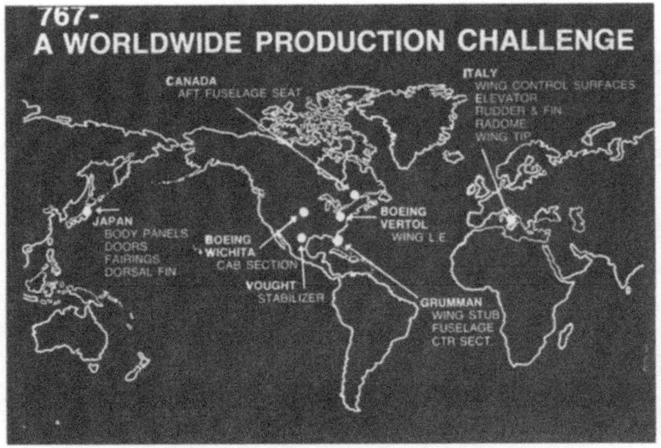

Figure 20

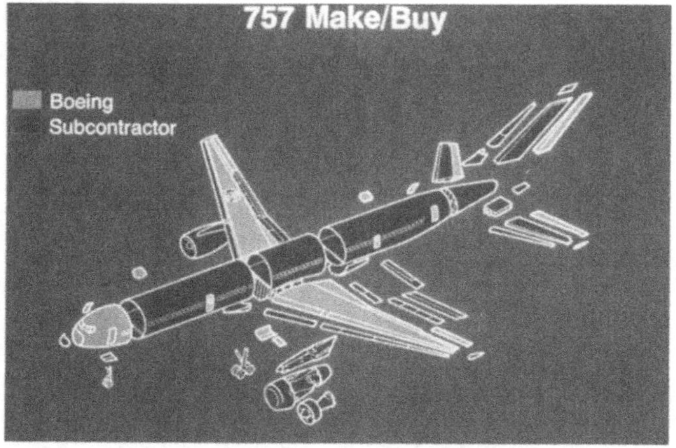

Figure 21

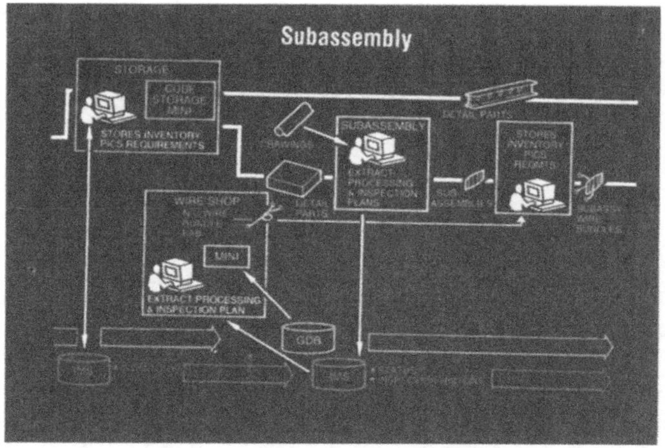

Figure 22

Figure 23

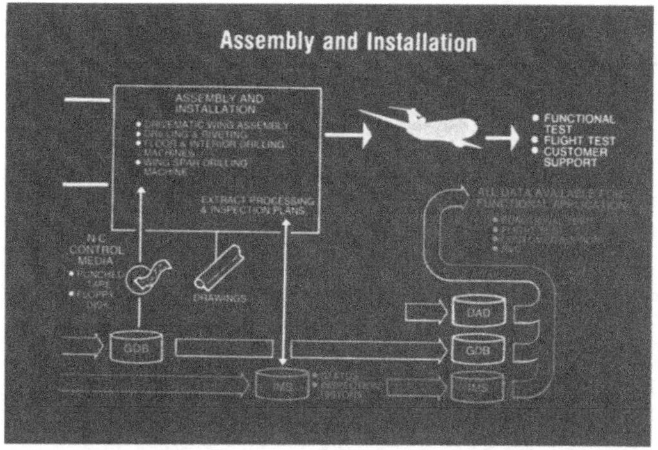

Figure 24

Figure 25

- o Flexibility: Remain flexible and adaptable.
- o Understanding Where the Savings Lie: One fallacy that has slowed the CAD/CAM movement more than any other is one that says engineering realizes savings in eliminating drafting personnel equal to the cost of the computing equipment. But a reduction in the work force is not where the savings lie. The savings from CAD/CAM must be realized in manufacturing and cannot come from drafting/design. Savings in engineering come later on down the line, in future programs. The introduction of computer-related techniques is a very expensive learning process until people begin to use such techniques as a normal tool.

SUGGESTED ADDITIONAL READING

Beeby, William D. May 1983. "The Heart of Integration: A Sound Data Base," *IEEE Spectrum*. New York.

Donaldson, C. 1981. *Tool Design*. McGraw-Hill Book Company, New York.

Florman, Samuel C. 1981. *Blaming Technology*. St. Martin's Press, New York.

Gerwin, Donal. March-April 1982, "Do's and Don'ts of Computerized Manufacturing", *Harvard Business Review*.

Harrington, Joseph, Jr. 1973. *Computer Integrated Manufacturing*. Industrial Press, Inc., New York.

Liewald, Michael H. & Philip R. Kennicott. May 1982. "Intersystem Data Transfer Via IGES", *IEEE Spectrum*. New York.

Litterer, Joseph A. 1973. *The Analysis of Organizations*. John Wiley & Sons, New York.

Moorhead, Jack, Editor. 1980. *Numerical Control Fundamentals*. Society of Manufacturing Engineers, Dearborn.

Krouse, John K., 1982. *What Every Engineer Should Know About CAD/CAM*. Marcel Dekker, Inc., New York.

BIOGRAPHY

Until the time of his retirement last July, Mr. William D. Beeby was the Director of Computer Systems, Engineering Division, Boeing Commercial Airplane Company and was responsible for the administration and direction of the computer activities within the Engineering Division and the interface with all other Divisions within the Company.

His assignments with Boeing have been varied: Chief Liaison Engineer–B-47; Assistant Chief Project Engineer–737; Manager of Production Engineering–Fabrication Division, where he was responsible for tool and production planning, tool design, production control, production order release, and all company NC activity.

He served as president of CASA/SME and on the Board of Directors of CAM-I and NCGA. Mr. Beeby won the 1981 American Machinist Award for his pioneering role in the development of computer applications that are leading to the integration of design and manufacturing. Having retired from the Boeing Co., he is now president of William Beeby & Associates, Inc., specializing in CAD/CAM consulting.

3

APPLICATIONS AND BENEFITS OF CAD/CAM DATA IN THE MANUFACTURING ENVIRONMENT

John H. Schulz
Director
Manufacturing Support Operations
McDonnell Aircraft Company
McDonnell Douglas Corporation
P.O. Box 516
Saint Louis, Missouri 63166 U.S.A.

Abstract

The impact of automation is being experienced throughout industry and in particular the aerospace industry. It is recognized that increased use of computers in engineering design requires an effective link to manufacturing for engineering data to be useful for automation of production processes and planning. To justify the resources required for continued CAD/CAM development, benefits have to be evident and near-term. The key to progressive computer applications in the engineering/manufacturing scenario is observing the benefits at the user level.

Presented in this discussion are CAM systems in operation at McDonnell Aircraft Company currently interfaced with engineering information at direct data flow points. Included are automated plotting of geometry for production parts, CAD tooling, CAD/CAM hydraulic tube fabrication, Graphic Numerical Control whereby engineering design data is used to create DNC machine tool programs through graphical representation, automated drilling of aircraft wing assemblies using CAD data, robotic drilling of acrylic transparencies and computer-aided technical illustrations for field service manuals. Also administrative CAM systems will be reviewed, such as Automated Drawing Parts List System, Common Bill of Material and Automated Planning. Improved productivity, improved reliability, reduced scrappage and other advantages of these CAD/CAM systems will be discussed.

1. INTRODUCTION

The impact of automation is being experienced throughout industry and in particular the aerospace industry. We, at McDonnell Aircraft, are recognizing that increased use of computers in engineering design requires an effective link to manufacturing for the engineering data to be useful in the automation of production planning and processes. Resources required for developing CAD/CAM systems must be justified with the benefits evident. We are finding the benefits of our progressive Engineering/Manufacturing computer applications to be downstream at the user level.

I am going to discuss some of McDonnell Aircraft's operating CAM systems which rely directly on information flowing from Engineering. Before discussing specifics, I believe it is helpful to understand the environment for which these CAM systems exist. (See Figure 1.)

MDC is comprised of four major divisions which occupy over 28,000,000 square feet around the world.

- The Douglas Aircraft Company is one of the world's main producers of commercial aircraft with facilities primarily on the west coast.

- o The McDonnell Douglas Astronautics Company manufactures missiles and spacecraft with its main facilities in Huntington Beach, California and Saint Louis, Missouri.

The McDonnell Douglas Automation Company is headquartered in St. Louis in one of the world's most advanced computer facilities. In addition to providing all types of data processing services internally, the Automation Company provides services to business, hospitals, schools, government, industry, and financial institutions. (See Figure 2.)

The McDonnell Aircraft Company–St. Louis is located adjacent to the St. Louis International Airport with facilities covering more than 500 acres. These facilities include 2,200,000 square feet of assembly operations and a 750,000 square foot Direct Numerical Control (DNC) machine shop. In addition, there is a new composite facility with 500,000 square feet and a new sheet metal fabrication building containing 300,000 square feet.

With McDonnell Aircraft Company (MCAIR) producing: (See Figure 3)

- o The Air Force F-15 air superiority fighter with 823 delivered and total orders of 1311. We are currently producing three per month and have reached a peak rate of twelve per month.
- o The Navy F-18 Hornet which replaced the A-4 and the F-4 in a dual role at nine per month with an expected peak of fifteen per month. There have been 123 delivered with total orders of 2572.
- o The AV-88 Harrier II V/STOL aircraft for the Marines current rate is one per month and expectations of six per month. Four have been delivered with total orders of 426.
- o In total, we are producing and delivering roughly one aircraft every other day.

One can now understand the advanced technology requirements which necessitate the use and development of CAD/CAM systems.

2. CONCEPT

Developing a Product Design Data System (PDDS) is the next major step towards reducing aircraft manufacturing cycle time and associated product costs. This concept has the design engineer providing manufacturing with a product definition in a form for automated production with the assembly fit verified prior to manufacture. Two-dimensional drawing reinterpretation, 3D reconstruction, and excessive assembly costs are eliminated. (See Figures 4, 5, 6 & 7.)

3. ENGINEERING INTERFACE

The CAD/CAM process at McDonnell typically begins with an engineer working at an interactive graphics terminal. Here, a product (such as an aircraft) or parts of a product are designed. After performing various calculations such as weight, volume, and strength on a real-time basis, a mathematical model of the product's geometry is loaded into a CADD (Computer Aided Design Drafting) data base. This information then becomes the vital link between design phase and the subsequent manufacturing of the end product. (See Figures 8, 9 & 10.)

An important key towards approaching totally integrated CAD/CAM Systems is the development of common data bases (see Figure 11). In many instances, the same basic data is required by many users who have in the past had to develop and maintain their own data sets. For example, the product geometry created by Engineering is now available to Planning, Tool Design, and the Production Shops. Common data bases help provide consistent and accurate data to all users, regardless of whether their application is centralized or distributive processing.

4. MANUFACTURING INTERFACE (See Figure 12)

By keeping in mind a "total integrated systems concept" when discussing Computer Aided Manufacturing, we will now move into a few of the CAM modules/applications in support of this concept.

The Automated Drawing Parts List System (ADPLS) (see Figure 13), using CADD geometry, creates and maintains an engineering "designed as" bill of material. In addition, ADPLS generates the parts lists portion of the engineering drawing and provides input to the common bill of material system.

An Automated Planning System was developed to assist planners in their effort to standardize fabrication operations (see Figures 14 & 15). By using interactive terminals and selecting appropriate transaction menus, the planners are able to improve their efficiency and eliminate the overlap and duplication associated with conventional methods when developing the planning for fabricating parts and assemblies.

Another CAM System now being used heavily at MCAIR by Production Planners is Graphic Numerical Control or GNC (see Figures 16 & 17). Using GNC, part programmers are able to access the same geometric data used by the Aircraft Designer to design an aircraft or sections of an aircraft. With this graphical representation, APT source statements are instantaneously created using routines such as AUTOPOCKET. Once the geometry of a pocket for a part has been defined, the APT statements necessary to completely machine the pocket are automatically created. After the APT statements have been created, the centerline data can then be plotted and superimposed on the part geometry for program verification (see Figure 18). When the programmer is satisfied the program is error-free, the part program is then stored in on-line files for use in fabricating parts on DNC machine tools.

RAPID is simply an acronym for Remote APT Processing via Interactive Devices (see Figure 19). This sytem basically supports on-line programming with inexpensive terminals and provides teleprocessing support for the DNC system. RAPID is a significant capability which allows programmers to access the on-line source part program files and make part program corrections or revisions interactively.

Another CAM application using the same mathematical model used by the design engineer and the production planner allows the Tool Designer to create the information required for fabricating and assembling various jigs, fixtures, and/or tools (see Figures 20 & 21). Then, much like in production planning, process control instructions are created that will be sent down to machine tools where tooling parts are cut out into flat patterns. These parts are then assembled in an "Erector Set" fashion using standard fasteners and brackets.

CAD/CAM technology is also being used in the tubing area where the CADD data used previously for Aircraft Design, Production Planning, and Tool Design is combined with additional information for the fabrication and inspection of the aircraft tubing (see Figures 22 & 23). In addition, this technology allows us to store tube characteristics in our master files and duplicate tubes which were previously developed under conventional mock-up methods.

I would now like to discuss three Process Control Systems being used in McDonnell's fabrication facility. These are DNC, IRS, and AC. Development of DNC, Direct Numerical Control, began in 1967 with the first machine being controlled directly from computers in 1970 (see Figure 24). This arrangement involved ten machine tools interfaced directly to an IBM 1800 computer. In the mid-seventies, the concept of DNC II was implemented, which involved eight machine tools interfacing directly to an IBM System 7 computer and then four System 7 computers would talk directly to the IBM 1800. This essentially allowed us to control 32 machine tools from an IBM 1800 and greatly enhanced our DNC capacity. More recently, however, our IBM 1800's have been replaced by a fully redundant Perkin-Elmer 3250 mini-computer system which we call our DNC Supervisor System. The full redundancy provides for a backup CPU, mirrored disks and the ability to switch any or all peripherals from one CPU to another.

The same software which is controlling 140 machine tools in St. Louis, with minor modifications, has been implemented on identical hardware in the Douglas Aircraft machine shop and shall be implemented in the near future at MDC divisions in Tulsa, Long Beach, and Titusville.

In addition to transmitting part program data down to machine tools, the DNC System receives over 4,000 management data entries per hour from the machine tools. These entries are generated through manual input with the majority being created automatically from sensors on the machine tool. Once this data is collected, it is then tabulated and reported on daily, weekly, and monthly hard-copy reports. There are also on-line CRTs available throughout the shop which enable direct and indirect production personnel to retrieve machine tool production information on a real-time basis 24 hours a day, seven days a week. Production information includes: part program run time, theoretical run time, machine tool down-and-why, tooling job, program tryout, etc. It should be noted that the on-line reporting allows quick detection and resolution of production problems as they occur. The DNC system controls a wide variety of machine types, including this 3-spindle 5-axis gantry profiler (see Figure 25). These gantries operate in pairs on 14 foot wide by 84 to 90 foot long beds and cost approximately two million dollars each. At MCAIR, we have 46 units in production, making this the largest complement in the world today.

This type of equipment is absolutely necessary when machining parts such as this F-15 bulkhead, with a raw forging weighing approximately 1200-1300 lbs. and costing $30,000/piece (see Figure 26).

The second process control system is our Improved Reliability System (IRS) (see Figure 27). This system is responsible for the detection and prevention of machine tool mis-positioning. Even with DNC and sophisticated multi-axis profilers, malfunctions stemming from a failing component, electromagnetic interference, or a power failure are still possibilities. IRS checks these malfunctions by comparing program data from the Perkin-Elmer 3250 and positional data from redundant feedback devices on the machine tool, on its own minicomputer. Anytime this on-going comparison detects a difference outside a given tolerance, a dynamic braking system stops any machine tool movement within five-thousandths of an inch. This type of system is relatively inexpensive insurance when machining complicated and expensive parts such as an F-15 bulkhead.

I would now like to discuss Adaptive Control (AC), our third process control system (see Figure 28). Adaptive Control basically monitors spindle deflection and then adjusts the machine tool feed rate accordingly, so that parts may be machined quickly and metal removal can be optimized.

Two axis Cutter Diameter Compensation (CDC) is currently being implemented on our 5-axis gantries in an effort to reduce part programming costs and to reduce and balance cutter inventories (see Figure 29). This is accomplished by inserting special command codes in the part program which the machine controller recognizes and then dynamically offsets the cutter path by the amount a reground cutter is undersize.

Composite manufacturing is receiving a lot of attention at McDonnell Aircraft Company these days (see Figure 30). With additional facilities just completed, the world's largest composite manufacturing facilities are approaching 500,000 sq. ft. with current production of composite material at 200,000 lbs/yr and peak rate of 500,000 lbs/yr expected. We have invested $70 million to date and will have invested $100 million by 1985 for the necessary facilities and equipment to manufacture composite material products. These modern facilities house a 1000 Watt CO_2 laser which cuts composite broadgoods at rates up to 600 in/min (see Figure 31), three Gerber knife cutters (see Figure 32), five 12/40 ft. autoclaves, eight ultrasonic C-scans, and 13 X-ray machines. Composite materials afford favorable strength characteristics while significantly reducing aircraft weight.

Automation in the assembly and checkout of fighter aircraft is also of keen interest at McDonnell Aircraft. We use 8,000 and 10,000 point wire (see Figure 33) analyzers to check the aircrafts' electronic circuits and equipment.

In a joint effort with the Navy, $7.5 million has been invested in an F-18 Automated Drill System (ADS) (see Fgiure 34). Using microprocessor based controllers, a Perkin-Elmer minicomputer, and a special four station wing drilling machine, holes are drilled in the substructure and wing skin with drilling programs received from the DNC System. Fasteners are than used to secure the skin to the substructure. An automated vision system has been incorporated to insure correct part matchup. This significantly reduces the fixtures and labor required in conventional wing drilling methods. A similar effort is now underway for an ADS on the AV-8B aircraft (see Figure 35). The proposed ADS will require expanding the 3-axis technology of the F-18 into 5-axis drilling for the Harrier.

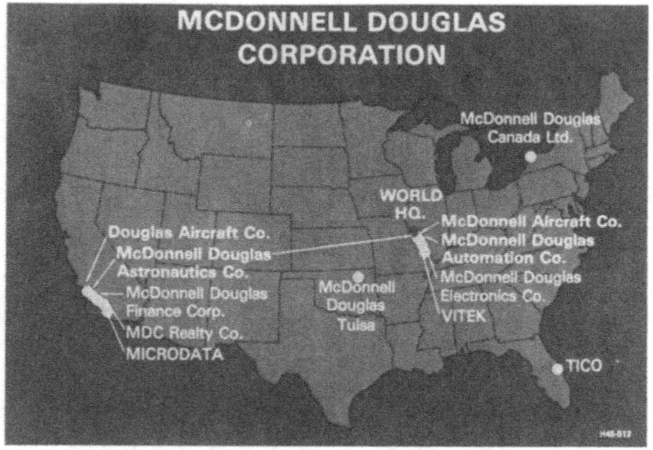

Figure 1

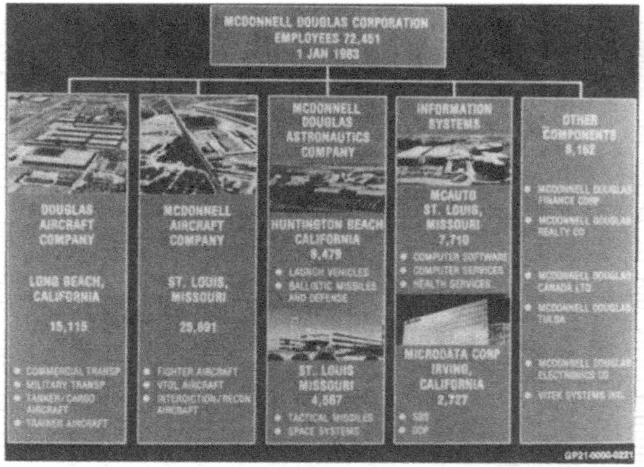

Figure 2

Figure 3

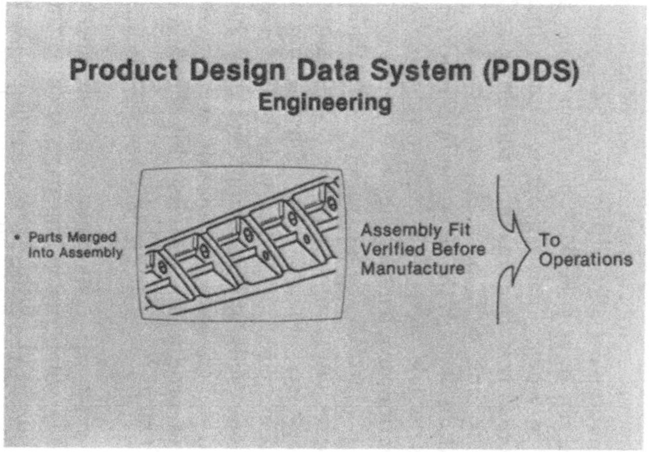

Figure 4

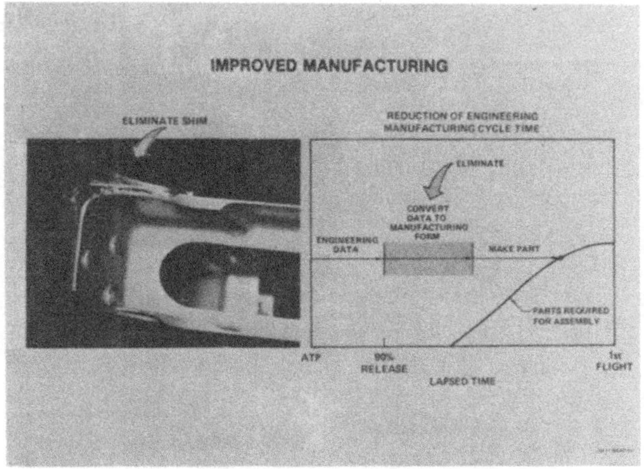

Figure 5

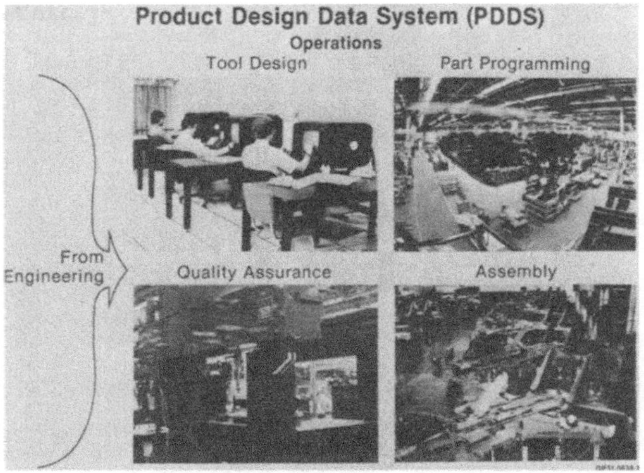

Figure 6

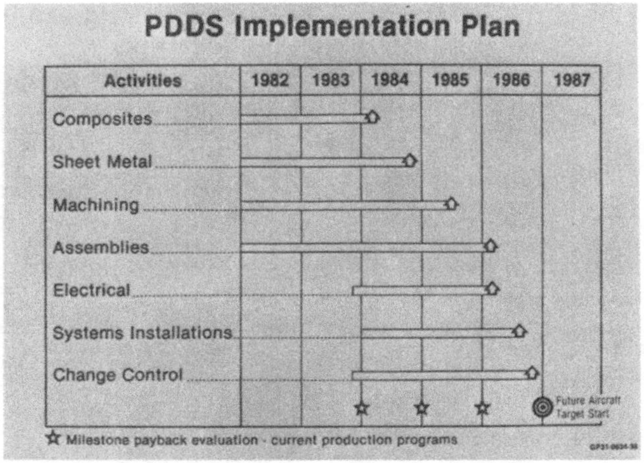

Figure 7

Figure 8

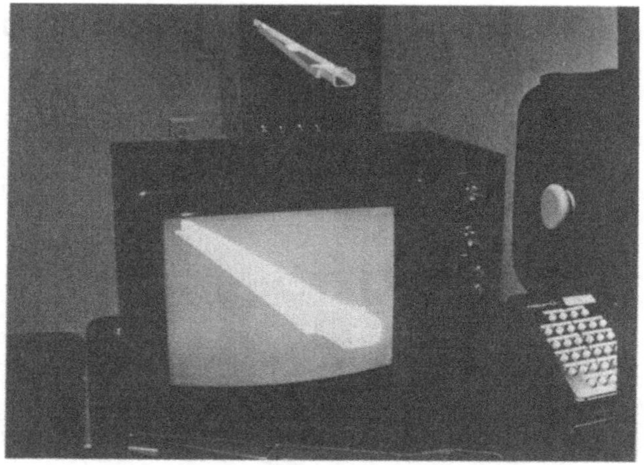

Figure 9

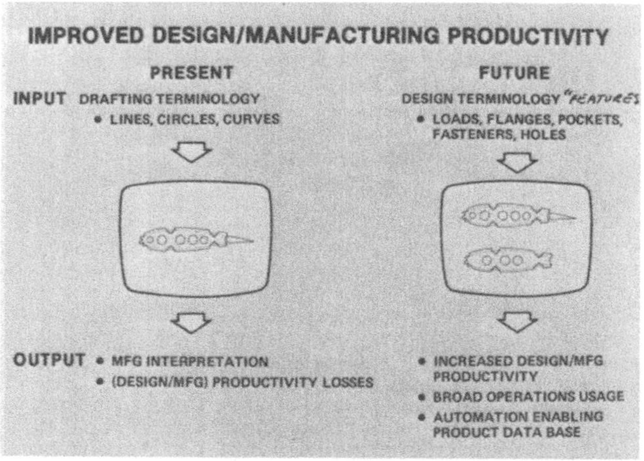

Figure 10

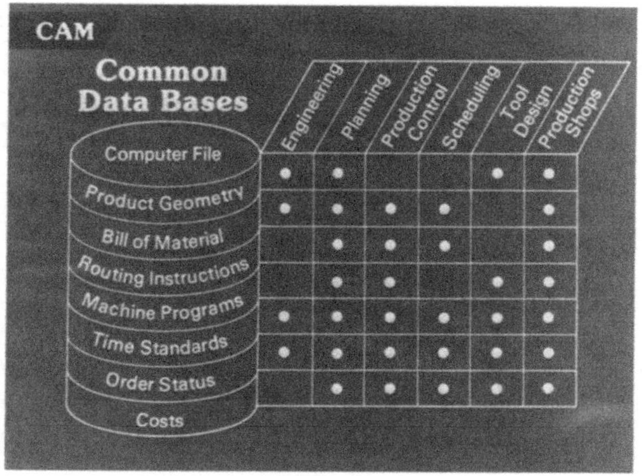

Figure 11

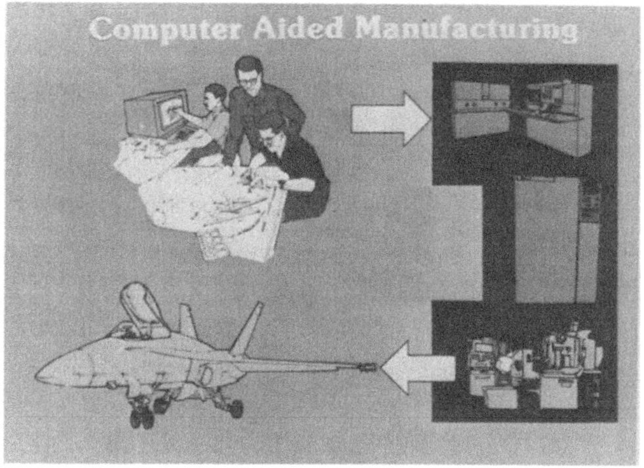

Figure 12

Figure 13

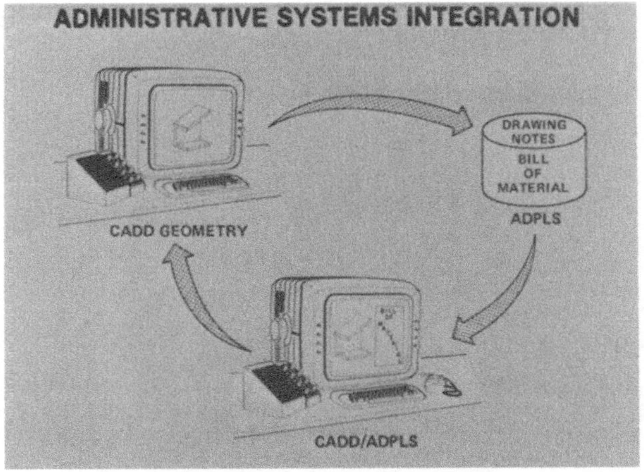

Figure 14

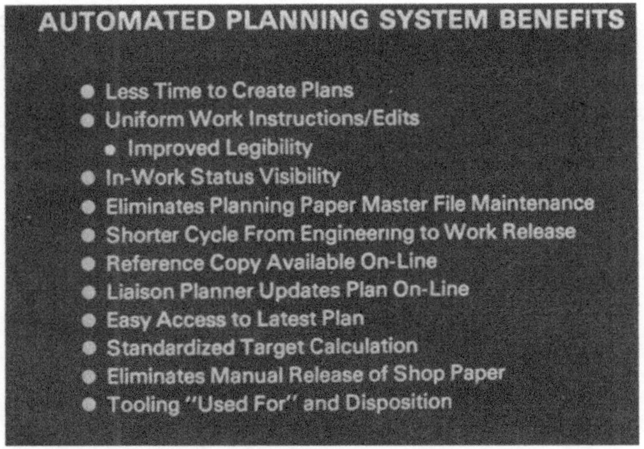

Figure 15

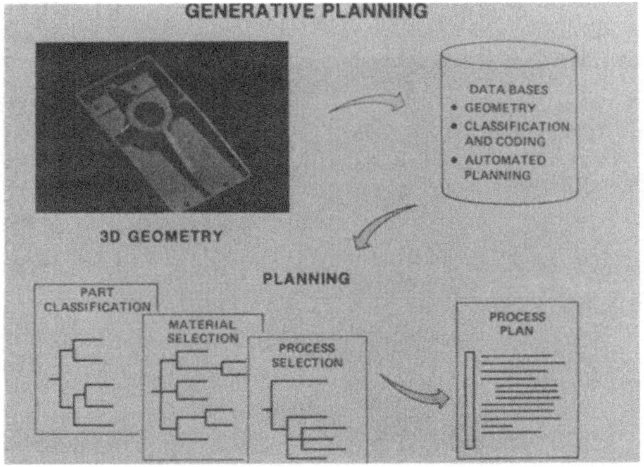

Figure 16

Figure 17

Figure 18

Figure 19

Figure 20

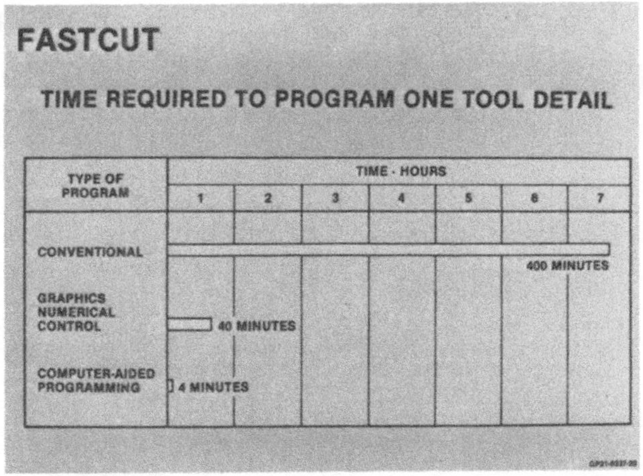

Figure 21

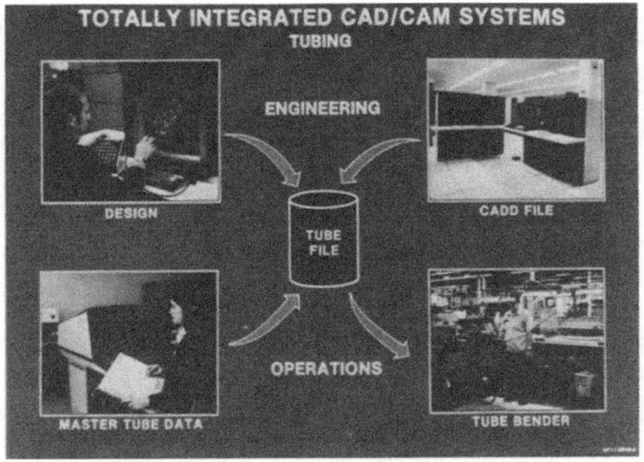

Figure 22

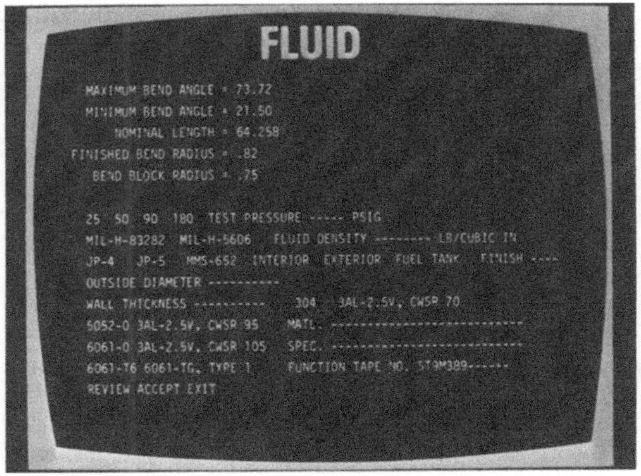

Figure 23

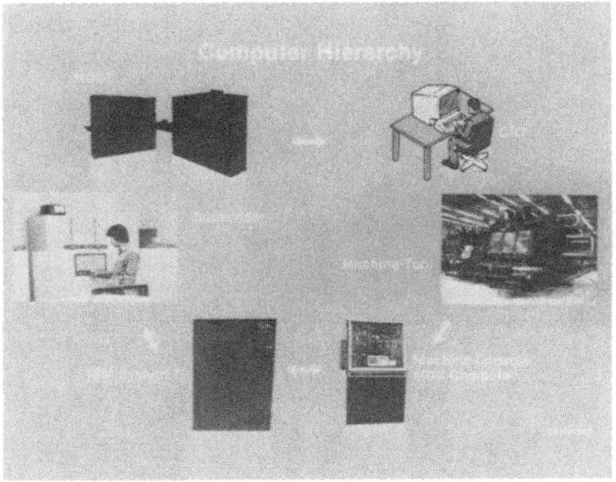

Figure 24

Figure 25

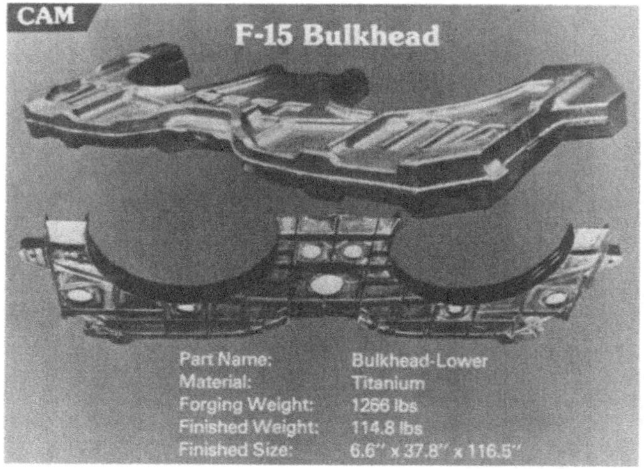

Figure 26

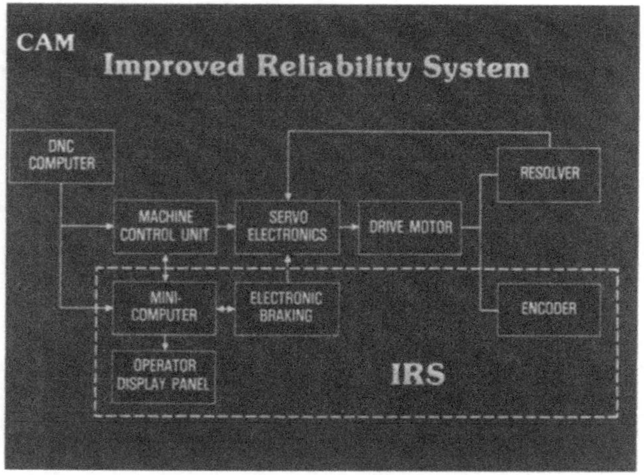

Figure 27

Figure 28

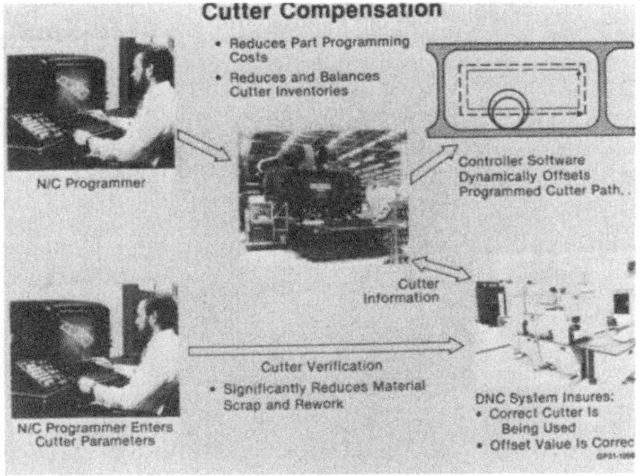

Figure 29

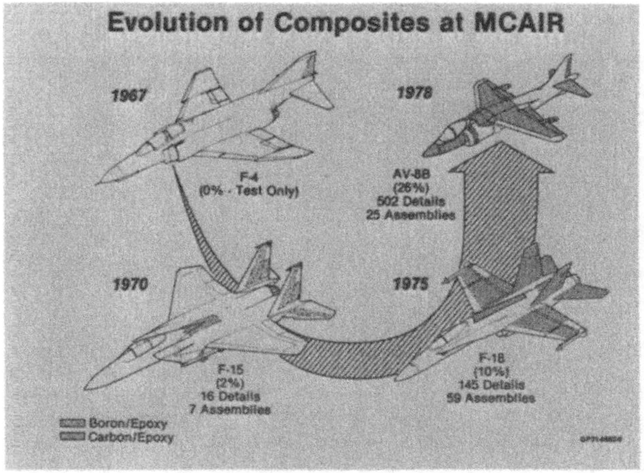

Figure 30

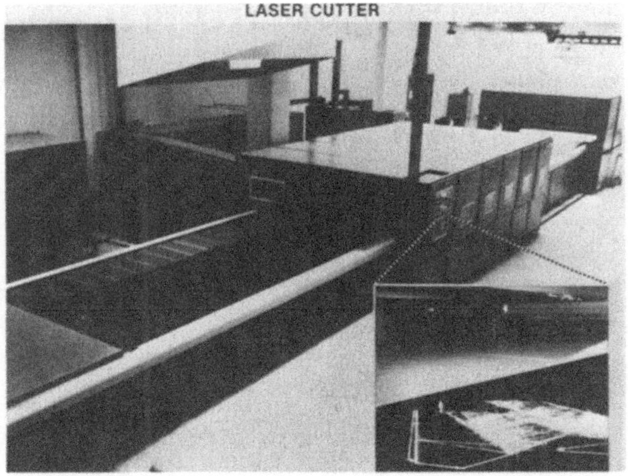

Figure 31

Figure 32

Figure 33

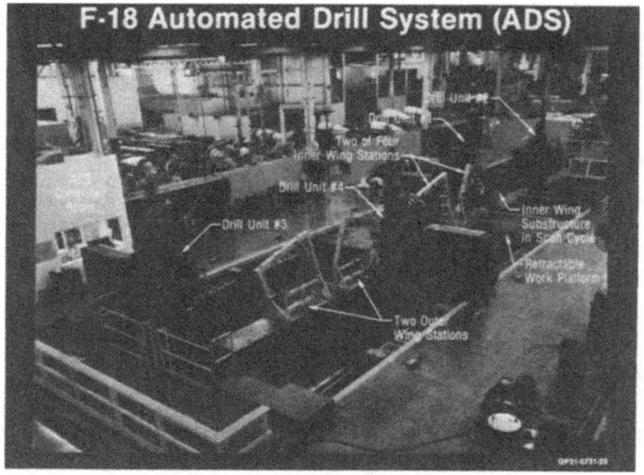

Figure 34

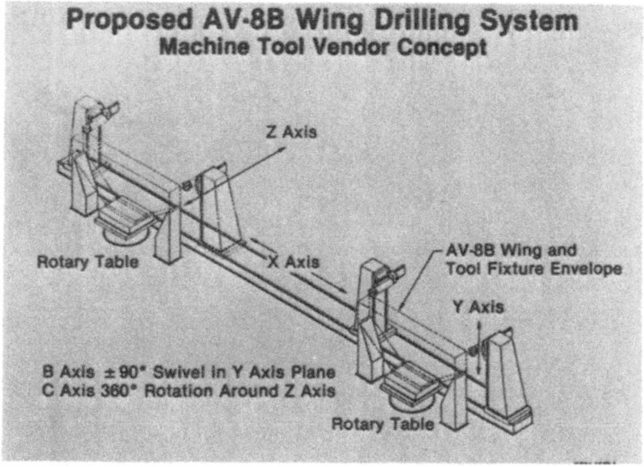

Figure 35

Figure 36

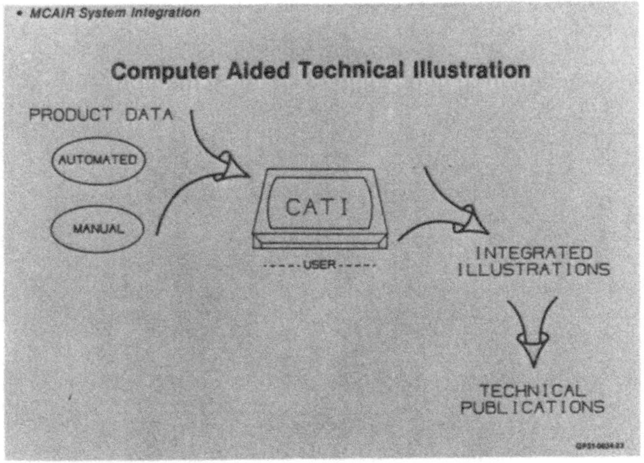

Figure 37

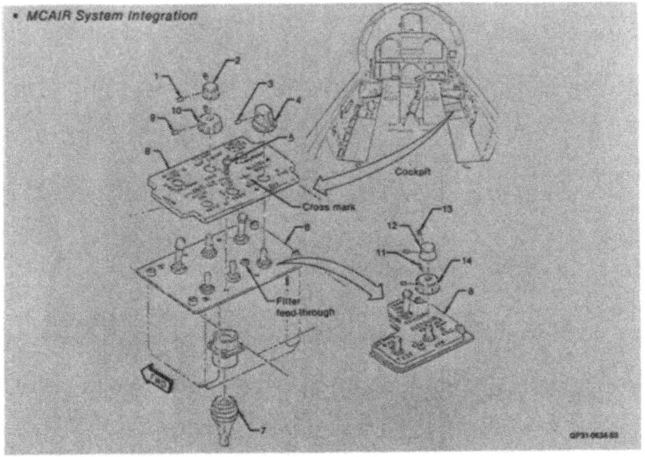

Figure 38

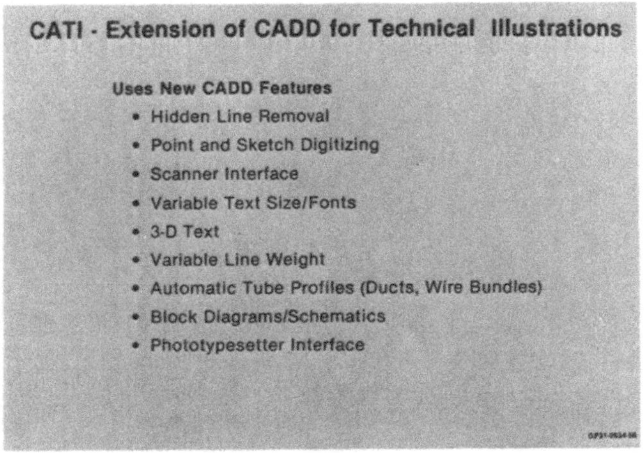

Figure 39

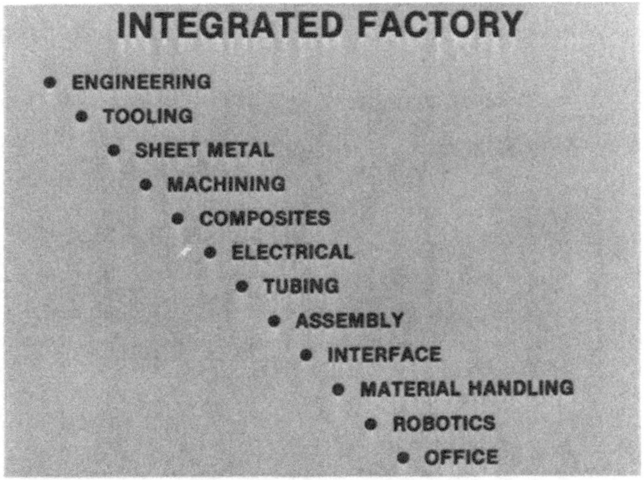

Figure 40

Figure 41

Robots are becoming an increasingly important part of industry today (see Figure 36). At MCAIR, a robot is being used in our Subassembly Shops as part of the windshield, canopy, and radome production cell for the F-15, F-18, and AV-8B. In addition, effort is being expended towards development of off-line robot programming and integrated "vision" systems.

Computer Aided Technical Illustration is a valuable extension of MCAIR's computer aided design/drafting system (see Figures 37, 38 & 39). By using the product data created by the Design Engineers, Product Support Engineers use CATI to integrate illustrations and automatically produce our technical publications.

5. NEW DEVELOPMENTS AND TRENDS

There are many new developments taking place in CAM technology at McDonnell Aircraft. I would like to mention one of these.

A Computer Integrated Factory (CIF) is now in the planning stages at McDonnell Aircraft (see Figure 40). To undertake a project of this magnitude requires top management commitment and investment. However, the benefits to be derived from CIF are numerous . . . they include:

- o Reduction/Elimination of Paperwork
- o Reduction of In-Process Inventories
- o Elimination of Data Redundancies
- o Automation of Manual Functions and many more

6. SUMMARY

I realize that in a very short time we have covered numerous CAM systems, but by no means all of them (see Figure 41). The systems and benefits I have mentioned are real and have proven McDonnell Aircraft a technology leader. To continue in this capacity, we are proposing a stronger integration of engineering and manufacturing.

Understand that additional costs are incurred with the engineer doing the lead-in process planning, but with significant gains being realized in the manufacturing arena with the elimination of human intervention and interpretation.

7. BIOGRAPHY

John H. Schulz is currently serving as Director of Manufacturing Support Operations of the McDonnell Aircraft Company, McDonnell Douglas Corporation, St. Louis, Missouri.

After one year as an Engineer with General Motors and two years of service in the U.S. Army, Mr. Schulz joined McDonnell Aircraft in 1957. He has demonstrated his technical and managerial ability starting as an Engineer in Methods Engineering, and progressing through Automation, Manager of Equipment & Process Engineering, Director of Industrial Engineering, to his present assignment as Director of Manufacturing Support Operations which includes Industrial Engineering, Production Planning, Scheduling & Control, Tooling and Manufacturing R & D.

Some of the more outstanding accomplishments in which he has been key include the design and implementation of:

- o A Direct Numerical Control System which operates over 100 machines, many of which are equippped with adaptive controls and failsafe systems.
- o The world's largest installation of 3- & 5-axis gantry profilers, valued in excess of $60,000,000.
- o A $70,000,000 composite material fabrication facility that includes automated laser cutting, honeycomb core carving, ultrasonic and X-ray inspection.
- o Automated production planning, part programming and tool design employing more than 100 interactive cathode ray tubes.

Mr. Schulz currently directs a team of 4,000 at McDonnell Aircraft. This team includes Engineers, Numerical Control Programmers, Scheduling/Production Specialists, Tool Designers and numerous Support Groups. He organized and served as Chairman of the Corporate Computer Aided Manufacturing Steering Group, reporting to the Vice President–Manufacturing. He holds B.S. Engineering and M.B.A. degrees from St. Louis University and has completed courses toward a Ph.D. in Engineering.

He has been actively involved in the AIA and CAM-I, Inc. and holds memberships in the AIIE, SME and NCS.

4

RELEASE CONTROL, CONFIGURATION MANAGEMENT, DESIGN RETRIEVAL, AND REMOTE VIEWING OF CAD GENERATED PRODUCT DEFINITION DRAWINGS

Robert E. Perri
Lockheed Missiles & Space Company
1111 Lockheed Way
Sunnyvale, California 94086 U.S.A.

Abstract

CAD usage at LMSC has been growing at a rapid rate. This rapid growth has created new challenges for effective utilization and control of this data. There is a need to make this mass of data more accessible to the doing organization and to provide for configuration control and management of the product definition. This presentation will describe some new software products developed at LMSC for release, control, and access of CAD generated design information for a large scale user population and explain why configuration management is considered essential to effective integration of CAD and CAM.

1. PRESENTATION

CADAM usage at Lockheed Missiles & Space Company (LMSC) has been growing dramatically. LMSC has almost 300 terminals supported by a centralized computer system. These terminals are located in 36 separate buildings, providing CADAM service to LMSC personnel throughout the bay area facility. This centralized CADAM system allows the 1100 active CADAM users to access drawing files from any of the 300 CADAM terminals. There are currently more than 100,000 drawings in the on-line files and over 150,000 drawings in the off-line files.

This explosion of CADAM usage has created new challenges in the effective utilization, control, and access of the CADAM data base. There is a need to make this mass of data more accessible to the manufacturing and support organizations. Also, it is essential to have a system of release control and configuration management of the computer generated product design information in order to implement effective integration of computer aided design with factory automation.

The recognition of these needs and requirements led to the development of the LMSC PREVIEW system. PREVIEW is a software product which interfaces to the CADAM data base and allows LMSC personnel to access and display CADAM data on the IBM 3279 administrative terminals. In addition, PREVIEW has incorporated software features to provide for release control and configuration management of the engineering data. The LMSC CADAM computer system is connected to the administrative computers via the multisystem network facility (MSNF). This network allows LMSC personnel to view the CADAM data base using any of the more than 800 administrative (IBM 3279) terminals in the network throught LMSC including remote sites such as Austin, Texas and Eastern Test Range in Florida.

To better understand the need for PREVIEW, it is necessary to review the limitation of the LMSC CADAM data base from the viewpoint of the end user of that data in manufacturing, inspection, etc. Some of these limitations are as follows:

LMSC CAD/CAM NETWORK

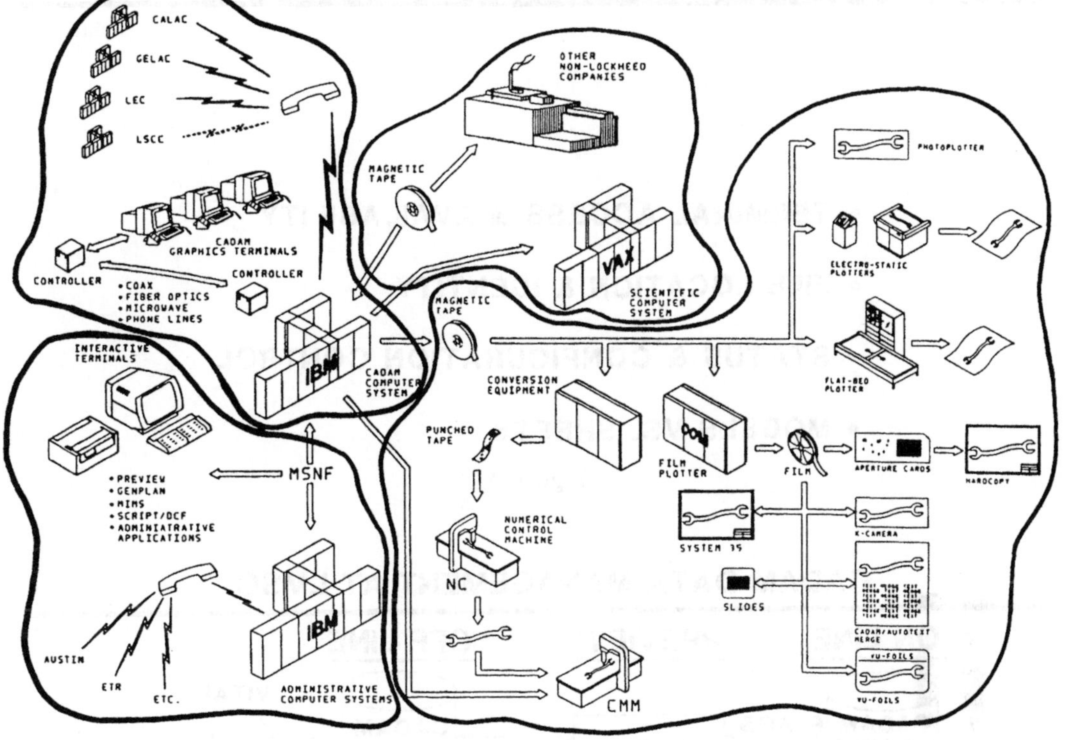

Figure 1

LMSC CADAM® USAGE

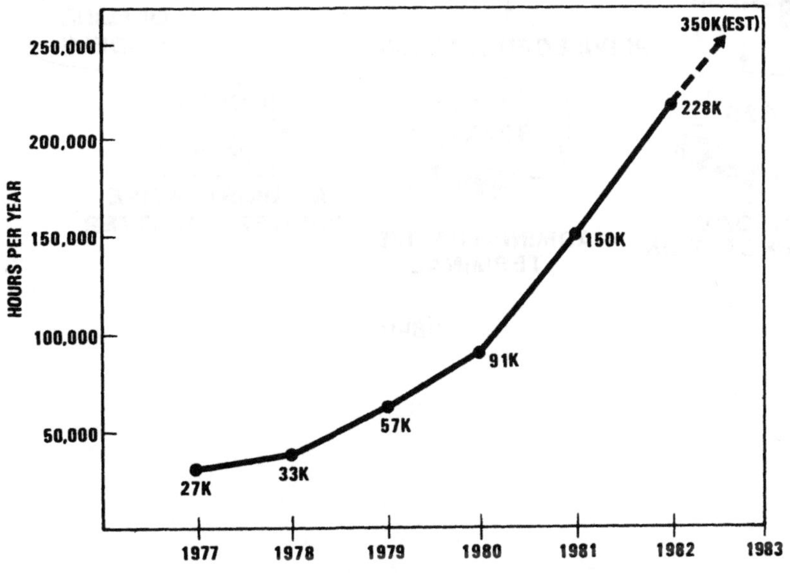

CADAM is a registered trademark of CADAM, Inc.

Figure 2

CADAM VIEWING LIMITATIONS

- TERMINAL ACCESS & AVAILABILITY
- FILE LOCATION & IDENTITY
- STATUS & CONFIGURATION CONTROL
- MODELS VS. SHEETS

Figure 3

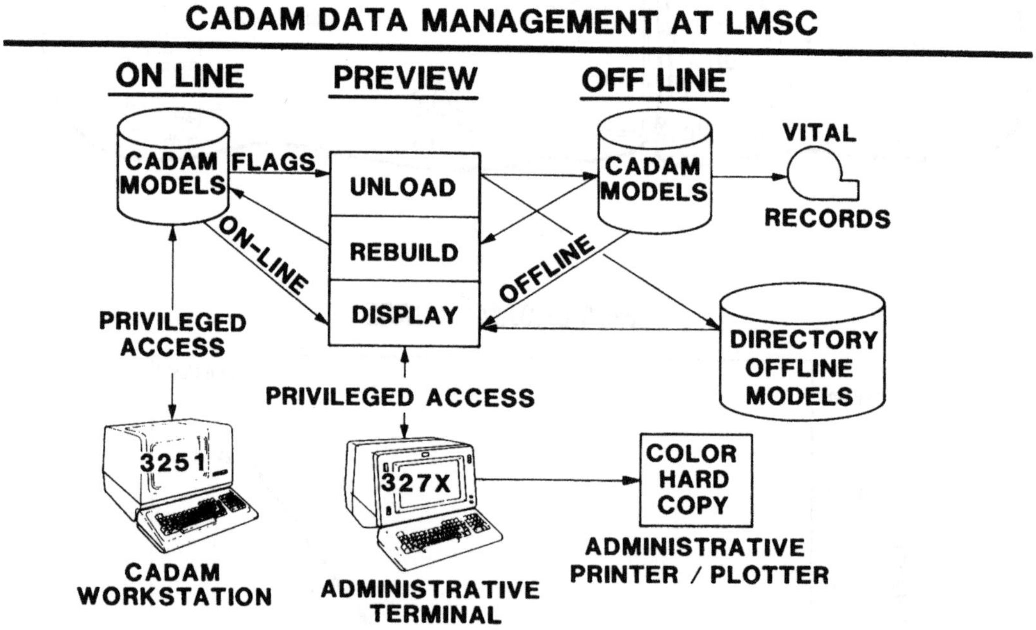

Figure 4

PREVIEW OBJECTIVES

- USE EXISTING EQUIPMENT
- SECURITY & ACCESS CONTROLS
- OFF-LINE FILE ACCESS
- RELEASE CONTROL & CONFIGURATION MANAGEMENT
- DESIGN RETRIEVAL
- DRAWING CHANGE COMPARE
- ANALYSIS

Figure 5

USES & APPLICATIONS

- CADAM OPERATOR ENHANCEMENTS
- DATA MANAGEMENT FUNCTIONS
- INFORMATION RESOURCE

Figure 6

- o CADAM Models vs. Drawings–
  The operator works with CADAM models overlaid along with a drawing format to make a complete drawing. The typical end user would want to be able to call and view complete drawing sheets.
- o File Location and Model Identity–
  The file structure is developed to meet the needs of the CADAM operator. LMSC currently has approximately 600 unique file identities and the end users would find it extremely difficult to locate the approximate CADAM drawing without knowing the file location and the models required to complete the drawing.
- o Status and Configuration Control–
  The CADAM on-line files are working files for the engineer and are continuously subject to change.
- o Terminal Access and Availability–
  The CADAM terminals are relatively expensive and generally not available to the untrained users of the drawing information.

Engineers create designs describing the definition of a product. The method of communicating this product definition is typically in the form of a drawing or specification. At the present time, computer-aided-design (CAD) systems provide a faster means to develop the designs. However, the output of this creative process is the generation of a paper drawing and the use of manual systems for release control and configuration management. With these limitations in mind, LMSC developed the PREVIEW software system to meet the following general objectives:

- o Be able to display complete CADAM drawings on currently available administrative terminals
- o Have the capability to view released CADAM drawings in the off-line files
- o Incorporate a design retrieval capability to access the released drawing by either part number or by key word in the drawing title text
- o Provide a method for release control and configuration management of the CADAM generated data
- o Utilize the color capability of the terminal to compare two revisions of the same drawing by overlaying the two revisions in different colors and comparing the differences.

PREVIEW was implemented into production in July 1983 and has received wide acclaim. The use of PREVIEW is rapidly changing the process by which end users access the drawing data and is playing a vital role in future plans for further factory automation. Configuration control and data access are two essential ingredients to effective implementation of factory automation in a large, diversified company like LMSC. PREVIEW is proving to be a giant leap forward toward meeting this goal.

2. BIOGRAPHY

Robert E. (Bob) Perri is manager of CAD/CAM services of Lockheed Missiles and Space Company in Sunnyvale, California. His department provides all the hardware and software support for a large CAD/CAM installation. Previously, he was Chief Draftsman of the Space Systems Division. He received his B.S. degree from San Jose State University and has been a part time instructor at SJSU teaching classes in Descriptive Geometry.

5

AUTOMATION TECHNOLOGY: THE KEY TO OUR PRODUCTIVE FUTURE

R. G. Freeman III
McDonnell Douglas Corporation
P. O. Box 516
Saint Louis, Missouri 63166

Abstract

In looking out to the next 15-20 years, it is quite apparent that Automation Technology will have the greatest influence on industry's ability to compete internationally. There are a number of critical areas where Automation Technology will impact the way we do business; particularly in training, logistics, factory and office automation, and in health care. To make sure this forecast comes true, we must continue to make advances in robotics, artificial intelligence, CAD/CAM modeling, and Computer Aided Instruction (CAI). This paper will attempt to forecast the contributions that Automation Technology will make to major industry initiatives to greatly improve our productivity and thereby our international competitive posture.

It is a pleasure to be back in the wonderful community of Monterey . . . particularly to talk about one of my favorite subjects: AUTOMATION TECHNOLOGY.

While I am not sure I can really define the term, I want to take the opportunity to do a little crystal balling with you in a different mode than the last time we talked. As I recall my previous discussions several years ago, I was primarily concerned at that time with the factory . . . CAD/CAM and office automation. This time I am going to talk on a broader agenda as I know that the specifics will be discussed by such able people as John Clancy, particularly in the area of CAD/CAM.

What I would like to talk to you about is a menu of subjects . . . how I see computer automation, automation technology and computer graphics contributing to a very bright and productive future. The subjects I propose to discuss are training and education, logistics, and health care. Then perhaps just a quick look to the future of robotics and artificial intelligence which are really subjects which I hope will be discussed in some depth at the Symposium.

However, I view my part in this Symposium as stimulating thought processes for the future, rather than dealing with specific technical issues. So let me deal first with the subject of TRAINING & EDUCATION.

I participated as a Committee Chairman in a DEFENSE SCIENCE BOARD STUDY in the summer of 1983 on training and training technology. While this study was primarily oriented to looking at the military problem, its conclusions are, I believe, applicable to the civilian sector as well. That is, that current technology provides unlimited opportunity in the TRAINING & EDUCATION field to substantially increase the productivity of our schools and universities as well as very definitely to enhance the retraining and continued updating of all categories of graduates. The DSB Study was an outgrowth of another study effort looking at whether our weapons systems were too complex for the personnel who were using them. The conclusion was that they are not, but that in many cases the training was inadequate and that technology offered a great opportunity to improve training effectiveness.

While the Information Services Industry has been concentrating on computer games, there appears to be limited emphasis on how we can incorporate all these new techniques in our training and educational systems. For example, in the aerospace industry, both commercial and military, extensive use is made of simulators to reduce training time or to act as a reinforcement of actual systems use. These systems are expensive . . . their technical requirements are as severe as the operating systems normally are. We can make great games by using simulators in engineering training; and that work I perceive to be still in its most embryonic state. The technologies of projection, of graphics, or simple devices are well within the current technology to provide assistance to students not only in the scientific and engineering fields but in gaming and case diagnostics. Several years ago, I purchased from awell-known water softener company a rather complex filter system which I was told was a "do-it-yourself" installation. While I have some technical capabilities, this one seemed to be more complex than I had anticipated. So when I called the company, they very promptly provided me with a video cassette which not only gave me complete installation details in the simplest lay language, but it also provided a series of "what-if-this-happens" type solutions as I went along, complete with graphics. While the do-it-yourself techniques have grown rapidly in the last decade, I have seen very little advantage of this taken, either in video cassettes or in video discs, except in industrial or military applications. It is therefore my opinion that we need to have, both in the military and the commercial side of training and education, research to establish criteria, methodology, and equipment for use at all levels of training. We also need demonstration projects for new training techniques so that data may be collected on performance and cost efficiencies.

Let me comment here that those data bases on performance and cost effectiveness are missing. For example, if you use a simulator as a training device, what level of proficiency does the student reach? at what rate does training or education decay? and to what levels do you return that student with refresher training? There is a press for the teaching of a variety of languages in our education system, and we most fortunately have been able to bring into this country a great number of foreign born who, while they understand the freedoms this great country offers, do not understand English. But you find the training of these immigrants is still handled in the same fashion. Yet there are new training technologies in computer graphics and automation technology to teach basic language and functional skills, while reducing dependence on detailed and difficult printed texts. This means increased R & D in voice recognition and synthesis, interactive display technology, personnel micro-processor training, etc., as well as application of VHSIC to the training problem.

System- or computer-specific software and graphics complicates operations and training and increases cost. We need to work on more generalized software designed for non-computer trained persons.

These recommendations were approved by the SecDef in February 1983. And while they are directed at the military, I believe they offer a major opportunity to improve the education and training in all levels of our civilian educational system; and thereby increase the quality of graduating personnel, which in turn will greatly increase the productivity of this country.

Now let me turn to the closely related subject of LOGISTICS/PRODUCT SUPPORT. There is no single area that more urgently needs to receive attention for future productivity than Product Support. We are long past the "Buyer Beware" philosophy of selling products. Logistics emphasis on the development of good distribution and transportation systems is important. But equally important are product diagnostics, product self-repair, built-in tests for standard commercial devices. This can be done at probably a significant reduction in the overall cost of a commercial warranty system. All the technologies are now here to allow the automation of our product support and logistics systems, such that central location with a master data base coupled to both wide & local area networks could be installed. We cannot only control our product support by providing instant diagnostics and graphics to go with it, but we can do it on most occasions without reference to anything but the data base. Take for example a refrigerator. If a customer had a failure in that refrigerator, a simple push-button device would tell him what part of the system was not operating. And by dialing a specific number and coding into the system, he could then receive graphics which would not only define the problems for him, provide wiring diagrams and solutions, inclusive of part numbers and the local distributor of parts, but also lead him through the repair job.

We are rapidly approaching, at least on the military side, the elimination of at least one level of maintenance because we can begin to build maintenance-free systems . . . very surely in the area of avionics and with complete diagnostics and mean-time between failure systems for hydraulics, pneumatic, and electro-mechanical devices. But here again the driving factor in making this happen is the integration of logistics and product support algorithms into the details of the original design during the CAD process so that the trade-off alternatives between product support, life cycle costs, and performance are made before a commitment to the final design. While the tools are with us, the culture still resists making this much needed change. The far reaching effect of imbedded diagnostic equipment on training is, I think, readily apparent. We now incorporate it in customer computer systems, in aircraft, and in some of the more expensive automobiles. But in my mind we have barely penetrated the R & D needed to modernize the commercial product support and military logistics systems to take advantage of current technology in preparation for the day when robotics and artificial intelligence will make our job even simpler and easier. I will discuss this point further later on.

The next area of importance I view for the future is HEALTH CARE. As a potential member of the Geriatric Set, I am and will remain very much concerned with the burgeoning cost of Health Care. And while it is normal to blame the physician for rising costs, the hospitals are major contributors to this problem. For those of you who are Systems Analysts and computer architects, I suggest making a trip through one of the hospitals for the application of automation and computer graphics. Extensive modifications to systems can be made with a high degree of cost effectiveness.

The use of computers in surgical techniques is a derivative of our CAD/CAM systems and is already a tested and usable technique. While the primary application of information technology has been in the finance area of the Health Care system, the application of cost economics and cost definition has not been emphasized as well. The same holds true of the basic patient care automated systems. The world of Health Care is changing as doctors band together to create their own patient care systems. The rise of Health Maintenance Organizations, Accident Care Centers, and a variety of means of serving the patient has cut the basic costs. The opportunities for application of new technology are growing. And the markets are continuing to grow. Again, I visualize the day when Home Health Care via computers will be the predominant way of delivering medical advice as opposed to many frustrating phone calls and trips to the doctor's office. As a matter of fact, in a recent physical that I took, the majority of my "vital" signs are characteristics, electrocardiogram information, etc., were piped into a central data base 1,000 miles away. Except for some rather obvious things which required the doctor to administer, the printer spit out the boundaries of a normal healthy individual, placed my results within those boundaries and wound up with a synthesized voice saying: "You are good for another year, Admiral!"

Here again the application of information systems technology accomplished with a few smart computer graphics capabilities we have, not only offers an opportunity to decrease the cost of Health Care Systems but to increase the availability of those systems to all economic levels.

These then are the three *primary* areas I wanted to discuss with you. They represent real breakthroughs for the application of automation technology, including computer graphics. The more obvious ones, such as automated cockpits in aircraft terrain following, ocean surveillance, ground mapping, distribution, automated factory, finance, and many other military and civilian applications all receive a great deal more attention than the three areas I have mentioned. But again, since my subject was a "PRODUCTIVE FUTURE", these three examples, I believe, will lead to a more productive future, since with training we will raise and maintain the level of training performance and thereby increase productivity. We will provide longer life to our equipment and services by Automated Product Support and Logistics, and thereby increase Productivity. Lastly, by the use of information services in computer graphics and health care, we will provide for a healthier environment and more Productive Person . . . and a substantial increase in the quality of life style.

ROBOTICS FOR ARTIFICIAL INTELLIGENCE

I think Professor Peter Wang would be a little unhappy with me if I didn't mention Robotics and Artificial Intelligence. The current wisdom says Robotics are here. And I would tend to agree with that, except that we have yet to solve the problem of sight and feel for our Robotics systems. Until that problem is solved, I believe we are still going to have widely split differences in the application of Robotics technology. Everyone can cite individual instances of excellent existing Robotics applications. But the problem is to make them productive and failure-free. And as I believe there has been some over-kill in the application of personal computers, . . . so there may well be in the application of Robotic technology. I tend to look at Robot technology a little like an erector set from which I may pick and choose the pieces I believe are most applicable to the areas of the office, the factory, or Services. In my opinion the Robot has its greatest use to gain repetitive quality and thus is a great contributor to productivity, if, in fact, you can achieve that. It does, however, create a new category of people who must care, maintain, and program our Robotics systems. Here again the definition of Robotic systems varies from a complete manufacturing system with a central computer servicing multiple robots to single arm actuators and sensors in a small machine or fabricating shop.

I believe, over the next decade, we will see major applications of Robotic Technology, but I also remember that for the last two decades we have been saying the same thing. They will have major application in removing the workforce from hazardous tasks and "dirty" jobs, for which the Robot can in all probability substitute. But I don't believe we will see the full application of Robot technology until we have solved the problem of expert systems and then on into Artificial Intelligence, and I don't believe we will see these systems within the next ten years.

While most of the progress in Artificial Intelligence has taken place in our universities, the application of this technology to the industrial process and the office is a good way off. It is a little like trying to automate decision theory. You very rarely make the same decision twice . . . and normally it does not take the same data base. That is the wonderful thing about the Human Mind which I think will defy any attempt to make it "Artificial".

In the two years since I last stood before this group, we have made amazing progress in a very rapidly changing industry. We have used and misused computers, computer graphics, expert systems and all the other areas of technology. On the positive side, we have created massive memory and computational capability in "chips", which five years ago was considered impossible. Galium Arsenide and other technologies offer even greater opportunities.

But let me close with a word of caution. The great storage capacity of modern chips, and most particularly custom chips, and the continued development of user friendly computer systems have, I think, lulled us into the belief that alll things can be done by computers. And I am sure there are some in this audience who believe they ultimately will. I remain with those who believe automation is a useful tool to increase our productivity and quality performance . . . but that we should approach the application of these technologies with some caution because the best computer in the world remains between our ears.

BIOGRAPHY

Rowland G. Freeman III has been serving as Staff Vice President–Strategic Planning for McDonnell Douglas Corporation since 1981. Prior to this, he served as the head of the General Services Administration, 1979-1980, and retired from the U.S. Navy as a Rear Admiral in 1979. RADM Freeman had a varied military career, including service as a Naval aviator, combat experience in World War II, Korea, and Vietnam; 5,000 flight hours in both reciprocating and jet aircraft; and five years of teaching at the graduate level. He also served as Commander, Naval Weapons Center, China Lake, California and was the Commandant of the Defense Systems Management College.

6

CREATING AN EDP ARCHITECTURE VIA COMPUTER GRAHICS – A TOOL FOR SYSTEMS INTEGRATION MANAGEMENT

Robert T. Bannon
Staff Supervisor
Systems Engineering–EDP Planning
AT&T – Communications
Bedminster, New Jersey 07921 U.S.A.

Abstract

The purpose of this paper is to explain the creation of an Electronic Data Processing Architecture using a general purpose computer graphics system. The EDP Architecture includes the interrelationships and data dependencies between EDP systems required for effective systems integration management. For purposes of illustration, this paper will use the functional relationships required to support telecommunications engineering.

By using computer graphics to define EDP architecture, the following advantages become evident: (1) Graphic mapping of interfaces inform EDP systems designers of user requirements. An example of this is the access of equipment requirements for determining floor space requirements. (2) Inform both users and upper management of proposed EDP design activity. (3) Capability to highlight major impact items. (4) Graphics structure allows the EDP architecture to clearly convey the maximum information required for a specific application function. (5) Dynamic graphics teleconferencing allows for immediate viewing from remote locations.

A suggested methodology of preparing an integrated EDP Architecture using Computer Graphics is described and can be applied to other technologies and industries.

1. INTRODUCTION

In a recent Booz-Allen study it was estimated that almost half the United States workforce is engaged in information processing today. There are a rapidly increasing number of companies and research institutions becoming involved in data manipulation, processing and information management. Because of this rapid growth, it is not uncommon to find that similar or redundant functions are being performed in various departments. In extreme cases, EDP systems within the same department may even store and manipulate the same data. This is caused by poor EDP system design, changing user requirements, parochial interests or proprietorship of information. Organizations do not want to disseminate their knowledge if this causes additional work or loss of total control of their system.

System integration architecture is a complex planning procedure that does not lend itself to analytical treatment of its effectiveness. Therefore, to plan system integration and an efficient data flow between systems, with a minimum of human interfaces and functional duplications, the establishment of an EDP Planning Organization is required. This group can reside within either the user or design organization. Its function is to create an integrated EDP System Architecture which depicts both the functional work flow and the interrelationship between systems. The basic thrust of the EDP Architecture is to provide an automated coordination and mechanization of interfaces for better data consistency and information management.

This paper discusses the use of computer aided design (CAD) systems to create the EDP system integration layout and to manipulate the EDP Architecture. By using the CAD system for these purposes, we are able to map the changes in information developed in one system and passed downstream to another. Also, we are able to convey, with a high degree of clarity, major impacts on the EDP system architecture. This in turn results in better system design for a new system or set of programs that will be integrated into the overall functional work flow.

2. CAD SYSTEM SELECTION CRITERIA

The selection of a computer graphics system to support the design of an EDP system architecture is similar in nature to selecting a graphics system to support either circuit board or building design.

Human interfaces with the graphics system are via a workstation. Today's workstations are ergonomically designed to relieve eye strain and body fatigue from working for prolonged periods. The operator creates, modifies, and refines the architectural layout interactively, viewing the emerging work on a graphic display. With a single command, the architectural designer can manipulate and change the proposed data flow, highlight impact areas in the output format or change the relational view.

A typical workstation will consist of a terminal (CRT), a hardcopy device and a variety of peripheral hardware that enhance the system's flexibility and responsiveness which the EDP architectural designers need.

To construct the EDP architecture and map functional relationships, you may choose to use either an existing in-house computer graphics system or opt for a new system. When using existing hardware and software, the initial economic outlay is less than that of a new system; however, there are some trade-offs that must be considered:

- o What is the volume of work already residing on existing graphics systems?
- o Is the processing environment a Mainframe or Mini-computer?
- o Is access time readily available, or is it on a special priority or allocation basis?
- o Is the workstation convenient for the architectural designer, such as in a common terminal area? Or will access be dependent on purchasing a new separate workstation?
- o Does the existing workstation design meet the user requirement of the architectural designer without modification? And if modifications are required, how extensive are they?
- o Is the software in a programming language supported by the potential designer organization? If software changes are required, does it require the vendor to make changes?
- o Was the graphics system designed to be menu driven only, or does it have the flexibility to allow the creation of special symbols, line manipulation and alpha-numeric text input and editing capabilities?
- o Is existing documentation sufficient to allow the existing CAD system to be manipulated for EDP Architectural integration?
- o What are the operational costs of the existing system?
- o How extensive is the EDP Architecture you intend to create? Does it support only one department, or does it encompass the entire EDP resources of your entire company?
- o Will your EDP Architecture include only the function and system level? Or will it branch into schemes and sub-schemes for functions and systems?

After investigating the capabilities and limitations of your existing CAD system, you may decide that it will meet your requirements or a majority of them effectively. If there are legitimate concerns about its efficiency and usefulness, you may want to investigate a new system.

It should be remembered that most CAD graphics systems are designed to allow interactive manipulation of symbol location and associated text. Business graphics systems and statistical graphic packages do not have the flexibility required for this aspect of EDP architectural design, therefore these systems are unsuited for an EDP architectural design application.

In investigating the purchase of a new graphics package to support the design of an EDP architecture, some additional considerations should be evaluated:

- o Total package price?
- o Cost/performance comparison?
- o Processor features: Micro-, Mini-, or Mainframe?
- o System capacity?
- o Interface compatibility?
- o Quantity of workstations system can support?
- o Expandability?
- o System performance specification?

3. CAD SELECTION

Based on an evaluation of our Engineering Department requirements, the time constraints associated with introducing a new graphics system and the versatility of an existing system, we chose to implement the EDP Network Systems Architecture on the Telephone Office Planning and Engineering System (TOPES).

TOPES is an interactive computer aided design (CAD) system that was developed by Bell Telephone Laboratories (BTL), Western Electric Company, and Long Lines personnel to assist office planning engineering in the layout and design of telecommunications and computer equipment offices. It is also used to size and layout administrative space requirements, support systems and building elements. TOPES is not a commercially available system. At the time of initial development, because of limited display capabilities, TOPES was developed around storage tube technology to accommodate the large number of drawing elements required to support space planning. However, with the current availability of high resolution refresh and raster scan terminals, TOPES has achieved device independence and can be accessed from a variety of terminals. The TOPES software utilizes Plot-10 software for its basic graphics subroutine package, and for object manipulation and transformations, however this software has been greatly enhanced by BTL. TOPES operates on multiple DEC-10 computers, therefore a large number of users can be supported without response time deterioration.

Time constraints will weigh heavily upon your evaluation and will influence your decision to use an existing in-house system, permanently or for an interim period, or opt to implement a new system.

4. EDP SYSTEMS ARCHITECTURE

EDP Systems Architectural Design is a discipline that evolved out of a combination of engineering and EDP program documentation criteria. It is based upon the CAD technology that evolved from schematic drawing layout for printed circuit boards and building design. At the same time, EDP architecture utilizes techniques associated with data flow and system documentation. The motivation for developing an EDP System Architecture is the creation and documentation of a logical model of system interrelationships for a functional or engineering technology. By using CAD graphical techniques, it enables engineers, users, analysts and designers to develop a concise picture of the various system dependencies and data access requirements existing between functional groupings of systems.

Ironically, system designers and programmers have relied on manual drafting techniques for system documentation, even while they were designing computer graphics systems. The use of a CAD system has eliminated this manual bottleneck in developing and publishing an EDP Systems Architecture. These noncreative and error-prone procedures can be costly and contribute to the untimeliness and inaccuracy of an EDP Architecture which must be a dynamic transition document.

Implementation of a CAD system in the EDP planning environment has eliminated much of the frustration incurred with making changes in a complex EDP System Architecture which contains a maze of engineering support systems that integrate functionally on information processing levels. Manual modifications that reflect system interconnection can be tedious, but with computer graphics the process will be expedited and the EDP Architecture will become a dynamic tool that can reflect transitional impacts from subschema level through functional level in a timely fashion. This allows the EDP Planning Organization to answer "what if" questions concerning downstream impact if the generation of certain data is eliminated or the format is modified.

With the old method of manually drawing system interrelationships, the input was usually a hand-drawn sketch prepared by the EDP planner, which was then redrawn into an understandable format that met corporate standards. This included the drawing of each system symbol separately, interconnection of these symbols based on the hand-drawn sketch, and finally annotating flow lines and placing an EDP system name within each symbol. Allowing for correction of errors and changes in subschema flows as projects become more defined, this process went through several reiterations before final distribution. This was a very time-consuming task, requiring constant interaction between system designers, EDP planning staff and the drafting department.

The use of the CAD system should not require the user to be familiar with a programming language, however our EDP planners have both an engineering and programming background. Therefore, the introduction of a CAD system into this process was an easy transition. The need for the development of a rough sketch and the drafting department translation into a corporate format that underwent multiple revisions was eliminated. Our CAD system has become the electronic scratch pad for the EDP planner, and consistently produces high-quality drawings.

The creation of an EDP System Architecture is usually created by repeating the following steps for functional engineering clusters:

- o Create a drawing grid or call up a standard grid from a symbol library, including border and legend generation.
- o Placement of individual system symbols in functional clusters on the grid. Symbols are drawn once and stored in a library from which they can be recalled from a menu or by a simple command.
- o Interconnection of system symbols. This is accomplished with line segments and arrows used to depict direction of interface flow or by use of connector symbols which are drawn adjacent to each system symbol. As a drawing convention, the connectors are used when the interconnection crosses three interface lines or five horizontal units. Each connector symbol uses an alphanumeric code, i.e., A or B2, and is recorded on an interface connector table for reference ease.
- o Reposition systems for functional cluster interface or interface with other systems.
- o Annotation or printing text on the EDP Systems Architecture.

5. CAD LIBRARY AND SYMBOL RETRIEVAL

The establishment of an EDP-oriented symbol library is one of the major factors contributing to the cost effectiveness and dynamic flexibility of using a CAD system to develop EDP architectural interrelationships. The storage of frequently used symbols greatly simplifies the preparation of the EDP Architecture drawing. These symbols may be simple, such as an arrowhead, or extremely complex, such as the EDP architectural symbol for a mechanized non-engineering interface process, which includes border shading. Once these symbols are created and stored in the library or database, the symbols can be recalled and incorporated into the drawing at the proper coordinate points. The method of retrieval is dependent upon the terminal hardware employed.

The use of a light-pen in conjunction with a menu is rapidly becoming synonymous with raster scan technology. This peripheral input device allows a symbol to be inserted into a drawing in seconds with just the touch of a button. Prior to 1982, systems based on raster terminal output were limited to farily simple displays because of limited screen resolution. However, availability of high resolution displays (up to 1024 x 1024) at reasonable prices have now allowed complex drawings to be presented.

With storage tubes, the initial drawing symbol is added on the drawing using a four-letter command after the terminal crosshairs are positioned at the desired X- and Y- coordinates. The symbol can then be copied at other drawing locations by repositioning the crosshairs to a new set of X- and Y- coordinates via the thumbwheel cursor controls or via a joystick and then issuing a single letter command (i.e., "c") to copy.

If refresh terminals are used, either method of symbol retrieval can be supported. This requires the light-pen and menu device to be incorporated into the workstation design.

6. CAD INPUT

The EDP Planner can design the architectural layout "on-line" or within a local, downloaded micro-processor which stores the drawing coordinates and associated symbols on either floppy or Winchester disk for block transmission to the host processor. The drawing input capabilities are inherent to the CAD system selected. Both modes of operation allow the system to respond immediately when compared to manual methods. A major advantage to the introduction of CAD systems into this area of EDP engineering is that as the architecture and associated data are being stored in the processor, visual feedback is drawn on the CRT which allows immediate on-line editing and rapid error correction.

When creating this type of drawing, many system symbols are the same; however, system titles are different which require the individual key entry of the associated text. Standard annotations, such as flow notes and text strings, can be stored in the library for recall. The text string can then be placed at the appropriate X- and Y- coordinates via the cursor/command sequences already described.

7. EDP ARCHITECTURE FUNCTIONAL GEOMETRY AND INTEGRATION

CAD systems use the computer's processing capabilities to eliminate many of the problems associated with the placement and interconnection of the various symbols utilized in an EDP Architecture drawing. However, for these features to be utilized, the EDP planner must establish a placement grid and a functional geometry.

Once the grid is established, the EDP planner functionally divides the grid for system placement. By functional division, it is meant that an area that serves an engineering discipline is reserved on the grid. For telecommunications engineering these functional disciplines might include:

- o Network Forecasting
- o Switching Planning
- o Equipment Planning
- o Inventory Control
- o Facility Planning
- o etc.

To establish an EDP System Architecture for other than telecommunications technology, substitute the appropriate engineering function or industrial responsibility for the functional division. An advantage of the functional division is that it allows the establishment of boundary coordinates for "zooming-in" on a portion of the overall Architecture. This allows small areas of the drawing to be copied in an enlarged state for purpose of clarity or if higher management's interests are focused on impact areas within a specific functional discipline.

Once the functional strategy and grid geometry are established, the various system symbols, i.e., network engineering symbols or non-network interfacing system symbols, are placed on the drawing in the manners previously described. The system symbol placement utilizes a strategy to minimize total length of the interface connection and limit line bridging once the systems are interconnected. This "mapping" technique is one of the most difficult for the EDP planner to apply. The planner must take into account the various drawing standards and physical constraints imposed by the functional grids and engineering clusters. Using the CAD system, the EDP planner must arrive at a functional placement that optimizes subsequent interface routes on the drawing without creating an undecipherable congestion. This is a matter of experience that may require multiple iterations, however, based on PC board and cartography applications, the use of a CAD system has a potential of a 5 to 1 time savings over manual drafting.

Some PC board graphics systems are equipped with an automatic router software which can be adapted for preparing the EDP Architecture. This is accomplished by assigning a location and zone to a symbol, similar to a gate assignment. For example, B7R3 to D4T1 would mean to connect the third right zone of the symbol in "B" vertical, seventh horizontal, to the first zone on top of the sumbol in "D" vertical, fourth horizontal. However, if your system is not equipped with an automatic router, and the input method is by terminal input, the interconnections are drawn by cursor control/command methodology.

If the CAD system selected for creating the EDP Systems Architecture utilizes either refresh or raster technology, there are several features that can be utilized to expedite drawing the interrelationships without abnormal congestion. Once the systems (symbol and system name) are placed on the drawing, a PC board design feature referred to as "RATS NEST" can be utilized. This feature allows the computer to determine interconnection paths and provides the EDP planner with a graphic representation of the proposed mapping. From this graphic display, the planner can determine whether a system must be moved to provide a cleaner interconnection route. At this point in the Architecture layout, the EDP planner can activate another PC board graphics design feature, "Rubber-banding." The planner places the graphics system into a dynamic mode which allows the EDP system and symbol to move or track across the screen and positions the symbol at potential locations. While tracking, the system interfaces stretch or bend while conforming to drawing standards, i.e., lines do not cross symbols but route around them. This gives the EDP planner the opportunity to select the optimum location. An adjunct standard that could be added to the "Rubber-banding" feature is the use of drawing connectors. If the interconnection line is forced to bridge across three or more system interconnections, or if the interconnection must pass through five or more horizontal grid areas, then the CAD system will default to the automatic selection of the next available connector, i.e., B2. The B2 connector would subsequently be placed adjacent to both the upstream and downstream systems and arrows would be drawn to show flow direction.

8. EDITING AND FLOW CHECKING

Throughout the layout process, it is necessary to perform various editing functions. The editing features are powerful commands that are incorporated in well designed CAD systems, whether they run on storage tube, refreshed or raster terminal. Interactive editing functions include commands such as layer or overlay discrimination and transfer. They may also include a glow command which allows the EDP planner to highlight major impact areas and systems. Another editing feature is the ability to change line type and weight, i.e., solid, dotted or dashed, heavy, etc. This feature conveys a high degree of clarity and information about a specific function. An example of this would utilize one line type and weight to show an interface between two IBM mainframe based systems, while another line type and weight could be used for a proposed interface between a system run on a DEC 10 and one run on a mini-computer.

Additional editing commands exist that control functions such as moves, deletions, text edits, and symbol and job renames.

If the EDP Systems Architecture is driven by an associated database, then a base exists to check the interconnections between systems. This can be approached either by manual inspection, or if your system is so equipped, by using the data extract process along with the default or "not-equal-to" linkage to print discrepancies.

At the same time, when a new system or interface between systems is added, the database design can flag additional changes and deletions and generate a CAD input report. For our telecommunications engineering systems database, we use RAMIS[R] as our data manager and report writer; however, other database management systems could be employed.

9. COLOR ENHANCEMENT

With the availability of high resolution raster terminals and monitors (1024 X 1024) like the Megatek and Hitachi units, color becomes an important feature for highlighting many different types of data. When creating an EDP Architecture, color can be used to show the telecommunications functional disciplines, i.e., equipment planning, system developmental status, EDP developmental agency, or even annual operational costs.

Low cost color plotters are available from vendors like Tektronix, Hewlett-Packard and

Nicolet-Zeta. These units can produce high-quality outputs that can be used by upper management as a decision making tool.

At the 1982 Autofact 4 show in Philadelphia, several companies including Tektronix and Megatek introduced low cost color ink jet copy units that can rapidly duplicate the raster screen presentation.

10. GRAPHICS TELECONFERENCING

If the CAD system is run on a mini-computer as its only processor, between four and eight remote terminals is the maximum configuration. This limitation is a function of access and processing capabilities. An advantage of a mainframe environment is that there can be a significantly larger number of available ports for individual users' access. Drawings can be downloaded to micro-processors located at the workstations for local graphics processing and transferred to the mainframe for drawing archival.

One of the most significant features of our CAD system is that of graphics teleconferencing from remote locations. Depending on port and conferencing bridge capabilities of the operating system and the CAD systems, it is not uncommon to have four to sixteen simultaneous remote conferees watching the same graphics output. By design, manipulation of the drawing symbols can be relinquished to remote locations, however the initiating drawing control ID maintains the ability to issue permanent overwrite or storage commands.

An advantage of graphics teleconferencing is that an EDP planner can be at a remote location and still make a presentation to upper management who may be at a remote management information and monitoring center.

An enhancement that we are investigating includes linking color code capabilities and subschema logic. The graphics teleconferencing when coupled with color capabilities and editing features allows "what-if" questions to be answered in an interactive mode. For example, if the long range (beyond five year) circuit input requirements from a conditioning system were eliminated, what would be the resulting impact on Space and Power Planning?

To answer this, the current path could be depicted in one color. Analysis of functional schema tasks performed in various Equipment Planning systems show that the circuits must be converted into orderable equipment quantities which are then converted to full bay equipment complements. The full bay complements are the basis for Space and Power Planning.

When the deleted input is shown, the subsequent functions glow or flash repetitively to indicate an engineering function jeopardy. The functional schema and subschema level must be searched to see if a modification or change can produce the required data. If it can, the "RATS NEST" feature can be activated to map the new interconnection path. When this is accomplished, the line type and glow frequency can be changed to indicate the answer to the "what-if" question. The EDP planner who has the drawing control ID will determine if the new input should become a permanent part of the EDP Architecture.

11. CONCLUSION

As in many technical fields, the design of Telecommunications EDP Systems Architecture requires multi-discipline expertise. This position requires a strong technical background in engineering functions as well as an understanding of EDP System design and development. The need for people with these special skills is consistently outpacing the personnel available. Examination of business opportunities are frequently postponed until an EDP Planner is available to address the EDP solution and incorporate it into the Architecture. This is even more apparent if the entire planning process is manual and computer graphics are not applied.

Therefore, it becomes essential that the EDP Planner's knowledge be recognized as a resource that can be enhanced with a CAD system and result in a productivity increase.

The advantages of introducing a CAD system into the EDP Planning environment are increasingly obvious. The CAD system rapidly produces an overall view of EDP systems interrelationships and functional dependencies.

The CAD system enables a human-computer synergism to exist. The system supports the creative abilities of the EDP Planner and the interaction of the two produces an Architecture that neither is capable of producing alone in a timely fashion. Clearly, a CAD system can produce significant time savings over manual methods.

Additionally, an up-to-date Architecture can provide a method for tracking EDP development projects and map their overall impacts. This keeps both the user and upper management informed of EDP activity.

Color graphics and editing features inherent to the CAD design allow the Planner to highlight major impact items and perform architectural "what-if" studies if a subschema level of detail is available.

Another benefit of using a CAD system to develop an EDP Architecture is the dynamic graphics teleconferencing capability which allows immediate viewing from remote locations.

CAD systems and applications are constantly being improved and enhanced and we have not yet discovered their technical limitations. The only suggested limitation is the human imagination. EDP Planners should be at the forefront of CAD system usage and take advantage of the mechanized capabilities they have created for other functions and technologies.

BIOGRAPHY

Robert T. Bannon is a Staff Supervisor, Systems Engineering – EDP Planning for AT&T – Communications. He is responsible for the development and administration of plans that direct the evolution of Engineering EDP Systems. These plans incorporate the introduction of new network technologies into the planning and engineering EDP Systems and provides for current and future communications and computer technologies.

Mr. Bannon is responsible for the introduction of Computer Graphics Applications into the planning and design engineering environment. This includes both CAD and Business Graphics. In addition, he is responsible for terminal and microprocessor evaluation. He has been involved in Computer Graphics since 1974 and EDP System Design and Programming since 1969. Previously he was a Space and Power Engineer for eleven years where he served as the Long Lines Project Manager and Systems Manager for various CAD and planning applications for Space and Power design engineering. He has also been involved in several Bell Telephone Laboratory and Western Electric engineering applications. In addition, he has been associated with several projects on automated digitization and character recognition, utilizing optical scanning techniques.

Mr. Bannon studied for a BSEE at Pennsylvania State University and is a former member of the Society of American Military Engineers. He is also a member of the NCGA, and has participated in the NCGA tutorial program. He has addressed the Computer Graphics Applications for Management and Productivity (CAMP '83) Symposium in Berlin, CADCON-EAST 83, the DPMA, and the American Right of Way Association. In addition, his work has been presented at the Computer Graphics 83 International Conference and Exhibition in London. He has been requested to participate in the Paris '84 Computer Graphics Exposition and the 1984 Harvard Computer Graphics Conference.

7

OPERATION OF JUKEBOX OPTICAL DISK OVER A LOCAL AREA NETWORK

Oliver E. Bessette and Michael E. Sullivan
Government Systems Division
RCA
Camden, New Jersey U.S.A.

Abstract

RCA's Argon Laser Jukebox programs are reviewed, including technical development and equipment manufacturing status. The oral presentation of this paper includes a video tape of the jukebox operation. The paper's focus is on RCA's recent work in coupling their first delivered jukebox via a 50-Mbps baseband Local Area Network (LAN) to a VAX/VMS host. The LAN interface was chosen because of the ease of interfacing with various host computers at data rates exceeding commercial data channel rates. Future work will use advanced LAN technology to interface higher data rate (>200 Mbps) optical disk memory systems to multiple and various users. These networks will include fiber-optic bus LAN technology.

1. INTRODUCTION

Optical disk memory technology has matured to the point of hardware development and testing of preproduction devices (1) (2) (3). Recently, engineering efforts have concentrated on the development of disk-drive components, such as lasers and mechanisms, and on media. Current work places more emphasis on the development of device controllers (4) and user access methods (5) (6). RCA started the development of digital optical disk memories in 1972 and currently has several Government contracts for media and equipment R&D and for deliverable jukebox devices with Local Area Network (LAN) interfaces.

Two preproduction 128-disk jukebox models are presently in various stages of manufacturing integration and checkout. A 50-Mbps jukebox will be delivered to RADC and a 100-Mbps jukebox will be delivered to NASA in mid summer of 1984. Each device can store 10^{13} bits (one million megabytes). The 50-Mbps system will be used to store large image files for the Air Force Multiple Image Exploitation System (MIES) at the Rome Air Development Center (RADC). These files will include reconnaissance imagery and mapping, cartographic and geodetic data. The 100-Mbps system will be used for a satellite imagery database in the NASA Data Systems Technology Program (DSTP) system.

The RADC jukebox will communicate data to the MIES via a 50-Mbps coaxial wire LAN. The NASA jukebox will communicate data to the DSTB via a 100-Mbps fiber-optic bus LAN. The selection of a LAN for the RADC system is discussed in the following paragraphs, including design and implementation aspects.

2. JUKEBOX LAN INTERFACE

Interfacing large capacity storage devices to computer storage and retrieval systems is not a straightforward task. A number of problems emerge during detailed design that often go unnoticed during preliminary design. These problems fall into three major categories:

(1) Operating System (OS) breakage
(2) OS software interface
(3) Electrical interface

Operating system breakage refers to the inability of computer operating systems to address the large amount of data stored in a jukebox system. The large number of tracks and sectors stored on a single optical disk exceeds the storage allocated for the maximum number of tracks and sectors within some operating systems. Some operating systems allot only 16 bits (65536) to the total number of sectors within a mounted disk. The optical disk exceeds this number by the 1638th track! A workaround to this problem involves writing a device driver that partitions the large physical disk into a number of smaller virtual disks with fewer physical sectors, each within the range permitted by the operating system.

Operating system interface drivers are difficult pieces of software to write. An interface driver is responsible for linking a physical input/output (I/O) device to the operating system so that the device can be used by application programs. They are difficult to write because the link between physical device operation and the operation expected by the operating system rarely fit. As an example, the operating system may picture the storage device in terms of ever-larger virtual sector numbers where the transition from the succeeding track is indicated by a single one-up increment. In this case, the software interface driver must convert from the virtual sector number to a physical track/sector number before making any data access. Of course, almost every computer system utilizes a different OS interface convention.

Electrical interface problems range from simple plug incompatibilities through complete signal incompatibility, exemplified by single-ended TTL interfacing to differential ECL.

In an effort to simplify optical disk interfacing, RCA engineers have selected the LAN technology to effect the optical disk/computer system (or other data source/sink) interface. The LAN technology offers four major advantages over conventional dedicated hard-wired interfaces:

(1) Device and operating system independence
(2) Potential for shared back-end processing
(3) High speed
(4) Available off-the-shelf for a large number of computer interfaces

Device and operating system independence comes from the use of the ISO model for system interconnection (Figure 1). Each layer shown in the figure performs only three functions: interfaces to the layers above and below it and communication with the peer layer in the distant host. The flow of data is from the topmost layer (applications layer) through the lower layers, to the physical connection mechanism, to the higher layers in the distant host. Control is strictly between protocol peers in this arrangement; for example, link level in host to link level in distant host. The independence comes from the agreement to the same peer protocol controls. In this way, a DEC obeying the ISO protocols can communicate with an IBM computer obeying the same ISO protocols. In our application, we would have the optical disk system itself appearing as a distant host and obeying the same protocols.

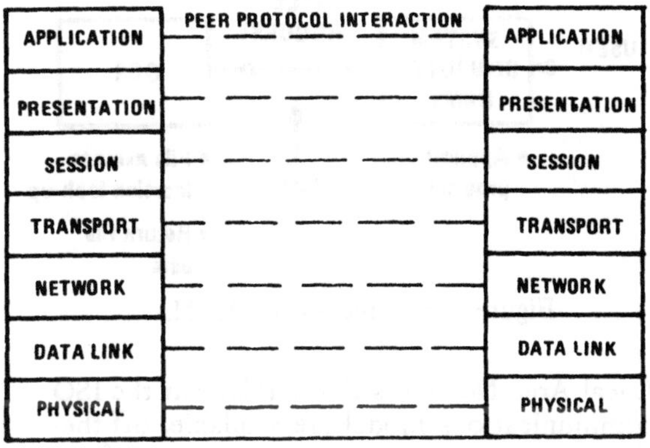

Figure 1. ISO communications model.

five years, he has directed engineering projects for image-compression and pattern-recognition hardware and optical disk controller architecture and software. He is a graduate of Worchester Polytechnic Institute, is a member of IEEE, has published seven technical papers, and has been awarded two patents in magnetic recording.

Michael Sullivan currently is Unit Manager, Software Engineering, RCA Government Communications Systems, Camden, New Jersey. He has 10 years of experience in software engineering, including the design and development of large software systems. For the past two years, he has directed the design and development of the control software for the RCA Optical Disk System. He is a former Naval Officer, a graduate of Duke University (BSEE, MSEE), an IEEE member, and has published two technical papers.

8

RASTER SCAN TECHNOLOGY OPENS NEW FRONTIERS

FOR THE MICROIMAGED ENGINEERING DRAWING

Gail N. Boyd
Versatec Inc,
Santa Clara, California

Abstract

Engineering drawings stored on aperture cards are scanned by a new 3500 element CCD (Charge Coupled Device). The image, now in digital form, is delivered to a printer/plotter for full size recovery. The scanner-printer link may be a standard data cable or the compressed image may be transmitted via modem, microwave, or satellite for delivery anywhere in the world.

1. INTRODUCTION

It can take hours, days or weeks to deliver engineering drawings. Courier services, mail and overnight delivery charges mount up. Manufacturing to the wrong revision level during lapse delivery time results in costly scrap. Satellite file maintenance consumes time and resources. There is an ever increasing need for a more time and cost efficient means of distributing the information stored on aperture cards.

2. APERTURE CARD SCANNER

ACRIS (Aperture Card Raster Input Scanner) is an aperture card digitizer specifically designed to output information according to Versatec standards and designated protocols. It is a raster scanning device in that it captures the image on the aperture card as a series of intensity modulated lines. Each line is sampled approximately 3500 times by a newly developed XEROX CCD (Charge Coupled Device).

2.1 SCANNING TECHNOLOGY

ACRIS combines electronic optical and mechanical technologies. The optics subsystem provides the illumination for the aperture card and directs that light to form an image of the card on the CCD. The CCD is the optical to electronic transducer providing, through electronic shifting of the sampled photo sensitive area, the raster scanning along the short axis of the card. Resolution is obtained by driving a precision lead screw slide mechanism with a stepping motor. For the image to be of high quality, the image motion must be smooth and precise. A microprocessor controller is dedicated to the generation of the special waveforms necessary to support the microstepping required for such precision.

2.2 IMAGE PROCESSING

The analog signal from the CCD is converted to a digital signal to which image processing algorithms are applied. Aperture cards with various background densities, such as those observed across the silver and diazo population, are accomodated through an automatic background adaptability feature which performs a line by line background measurement and corrects for observed trends. A built in threshold adjuster sets the conversion level for a black or white decision. The operator may change the threshold level to one of seven settings on the interface panel in order to improve fine line detectability or remove background noise. Pixel interpolation and selection algorithms guarantee full magnification to original drawing sizes A to E and A4 to A0 at 200 x 200 DPI resolution.

2.3 PRINTING

In a hardwired configuration the raster output from ACRIS is in a parallel format compatible with the Versatec electrostatic plotter. The plotter may range in width from eleven inches to thirty-six inches. ACRIS is capable of scaling the output according to the print width selected on the user interface. A certain amount of versatility accompanies electrostatic plotting: the scanner may be located up to 1000 feet away from the plotter; it may be multiplexed to a plotter already in use as a computer peripheral; and there is a wide selection of output media available including film and vellum.

2.4 TRANSMITTING

When the aperture cards and the scanner are miles away from the distribution point, ACRIS with the remote option board installed will deliver the print in minutes over a variety of carriers. The raster data is compressed using a two-dimensional run length encoding technique. The compressed raster is sent out over RS232 or 449 using a subset of the SDLC protocol. A remote receiver decompresses the data and formats it for the plotter. Line speeds which will result in quality output and are supported by ACRIS range from 4800 baud to 56K baud. Transmission times will depend on the line speed and the degree of compression achieved. Aperture cards present a unique situation in that the filming and mounting process contributes to the degree to which the image remains orthogonal and thereby affects the efficeincy of the compression algorithm. Engineering drawings are expected to compress from 10:1 to 30:1 depending on the complexity of the image.

The ability to transmit drawings addresses a number of time consuming and costly situations. Satellite files of aperture cards are maintained in order to increase availability of the most recent revisions of drawings. Even with this arrangement there may be a two to seven day delay between actual revision and availability of the current card. The cost and efficiency of maintaining many separate files as opposed to one central file is addressed by a scanning unit placed in the central file linked to the various sites by communication lines. Not only are satellite files unnecessary but the most current revision level is guaranteed since the time consuming distribution process has been eliminated. In manufacturing a considerable amount of work is scrapped each year at losses amounting to millions of dollars as a result of working to the wrong revision level. The ability to receive a drawing in the manufacturing plant in minutes regardless of it's location offers a solution to this costly problem.

Another benefit of digitizing and transmitting aperture cards is the level of response during real time failure situations. Facilities piping and electrical layouts may fall into this category. Secured areas such as nuclear power plants need access to information quickly which may be stored outside the secured perimiter or in another location. Distributing drawings to another county can be a major source of frustration. It typically can take several weeks for drawings to reach for example, Japan or the Latin American countries. It would not be unusual that in that time the drawings were returned rather than delivered. ACRIS technology offers the solution to the distribution problem. It couples the advantages of digital data with the cost efficiency of storing on microfilm. In addition to providing flexibility and responsiveness, the ACRIS product has created some exciting potential expansions for the microimaged data base.

2.5 NEW FRONTIERS

ACRIS creates a digital representation of the aperture card image. In this form the data can be displayed on a compatible raster terminal for viewing. A compatible interactive graphics terminal will provide editing or redrawing capabilities. The data, in it's compressed form, would lend itself to loading a digital data base. Today this would be magnetic storage but consider the tremendous development efforts in cost efficient mans storage, particularly the Optical Data Disk. More elusive but just as exciting is the effort to vectorize the raster scanned image and convert it to a form recognizable by CAD systems.

Looking ahead to new frontiers - drawings currently on microfilm might some day share all the benefits and advantages the digital world has to offer.

3. BIOGRAPHY

Ms. Boyd of Versatec Inc, A Xerox company, is the Product Manager of the Aperture Card Raster Input Scanner (ACRIS) Program. She is deeply involved in products representing pioneer breakthrough capability in raster scan technology, electrostatic plotting, image enhancement done on a digital basis as well as image-delivery of engineering drawings over a variety of carriers. Ms. Boyd received her BA in Biology from the University of Northern Iowa and was awarded her MBA by the University of Santa Clara.

9

COMPUTER-AIDED TECHNICAL PUBLICATIONS

Deborah Gail Brown
Ford Aerospace and Communications
Ford Motor Company
307 19th Street
Huntington Beach, California 92648 U.S.A.

Abstract

All phases of the engineering design and manufacturing processing have been either monitored by or produced on a computer for over twenty years. However the last phase of production, technical manuals, has remained a manual process until recently. Portions of the technical manual process were automated some time ago. Several vendors have produced systems which attempt to integrate the automated portions as well as provide enhancements. A true understanding of the technical manuals production cycle has not been developed by the vendors. Herein lies the crux of the problem with developing a superb system.

Several months of research culminated with the result that a "good" technical publications system does not currently exist. The author, therefore, proposes a system which satisfies the requirements of Technical Publications.

1. INTRODUCTION

It is the purpose of this paper to discuss the automation of a technical publications group involved in the production of manuals for a high technology corporation. Several companies advertise "complete technical publications systems" which provide both a drawing capability and text handling features. The layout of a manual page requires placing text and graphics together on the same page. This process may be done automatically on an interactive graphics display terminal or manually by cutting text apart and pasting it onto the artwork. It is, however, the author's contention that the application is too new and the understanding too infantile to produce a "good" technical publications system.

1.1 GRAPHICAL APPLICATION AREAS

The computing world has known graphics for several decades in the form of CAD/CAM, Computer Aided Design/Computer Aided Manufacturing. The aerospace and automotive industries were among the first to employ CAD/CAM systems in an effort to cut costs and increase productivity. Large computers were required to drive the design terminals. The initial investment was enormous but these companies thought the investment would soon pay off.

Today the CAD/CAM market is no longer monopolized by one large company as it once was. CAD/CAM systems may now be purchased as turnkey systems. The increased computing power of the current generation of computers permits the execution of an entire CAD system on a microcomputer. The decreasing cost of both hardware and software allows smaller companies to use graphics in their design work. CAD/CAM is becoming a necessary tool for any design-oriented company to remain competitive.

Computer graphics has made it into the entertainment field by way of arcade games and animation systems. In fact, computer graphics are being integrated into every major aspect of business. Technical Publications has a history of automating isolated functions as the equipment has become available. As Larry Letteny from Eastman Kodak stated,

> "We have so few functionally integrated businesses in the industry. Usually one guy prints, another guy does separations and so on." (4)

For instance, an optical scanner takes a page composed of drawings and/or text and transforms it into digital data. It is then fed to another manufacturer's computer-driven photocopier. The art is produced manually as well as the layout of the page. Several manufacturers of turnkey CAD/CAM systems have recently begun marketing a "Technical Publications" system with limited success. The CAD/CAM industry has always required high quality in both the design and output stages due to the nature of their work. The illustrations of technical manuals are closely related to the CAD/CAM field. Each one is a compilation of several engineering blueprints which produces a drawing similar to Figure 1.

The question may be asked, "Why, if the publication of technical manuals is a rehash of some CAD/CAM application or drawing, has it taken so long to market a computer graphics system for technical publications?" One answer might well be that CAD/CAM manufacturers were so busy trying to keep up with the CAD/CAM industry that there was no time to further modify the system to encompass another application area. It was a full-time job to handle new computers, terminals, output devices and user enhancements. Another deterrent was cost. Until recently, a technical publications group could not afford the equipment for a design system. The most that could be hoped for was a design station that was attached to the engineering CAD system.

The U.S. Government specifications for manuals now include a requirement for 75% art and 25% text as opposed to the old format of 75% text and 25% art (3). Government contractors are faced with a sizeable increase of artwork for their manuals, the costs of which are being used to justify a graphics system for the technical publications group.

1.2 PROBLEM AREAS

Only very motivated illustrators would attack learning a CAD system. The terminology and construction techniques are aimed at engineers and draftsmen, not artists. For example, engineers and draftsmen deal in very real terms. A line must be 10 inches long and at a 60 degree angle to another line which is 3.765 inches long, at a 31.05 degree angle to the origin. Illustrators, however, are concerned with how the two lines appear. Are they in the proper viewing perspective? Do they look right? A draftsman would construct the two lines by keying in lengths and angles while an illustrator would be more comfortable constructing the lines by freehand drawing methods. The CAD/CAM system is also capable of many analysis functions which engineers need but illustrators find unnecessary and not cost effective.

Furthermore, CAD/CAM companies do not understand what is required for technical manuals. It has been apparent that the end drawings resembled an engineering drawing but how it was conceived was another task. Another problem is the large amount of text associated with a technical manual. Drawings are not accompanied by instructions on how to assemble/disassemble a part or piece of machinery. The tools required to perform the task are included, as well as a general description of the part. The manual may also include trouble-shooting instructions and numerous other types of verbage. Engineering drawings, on the other hand, contain minimal text that is typed in by the operator or generated by one of the analysis functions.

1.3 TECHNICAL PUBLICATIONS SYSTEM REQUIREMENTS

The first problem is to understand the requirements of a Technical Publications system and the second is to solve the problems associated with satisfying these requirements. An understanding of the artistic drawing processes must be developed. This affects the interface to both the drawing and the system functions. A survey of illustrators produced a set of requirements aimed at the drawing capabilities. Four vendors were then surveyed in an effort to discover whether these requirements have been met. The results illustrate the lack of understanding prevalent among CAD/CAM vendors. Simple drawing features are beyond their comprehension although each CAD/CAM/Technical Publications system is capable of numerous other functions such as stress analysis, numeric control instructions and generation of tooling paths.

The driving force behind the new technical publications systems seems to be the need for a new marketplace. The CAD/CAM market is being saturated; major engineering firms have made their investments and only maintenance and additional hardware are required. Technical Publications, on the other hand, is a virtually unexplored area. Although the hardware requirements are the same as for CAD/CAM applications, additional requisites include a new user interface with such features as automatic hidden line removal and rotation of isometrics, interfaces for a word processor and a phototypesetter, and the removal of all analysis programs. Due to the many similarities and the need to access current engineering drawings, CAD/CAM vendors are attempting to quickly capture the new marketplace by modifying their software to accommodate the new demands. These systems, however, are overdesigned and undersimplified.

1.4 TECHNICAL PUBLICATION MODEL

Manual construction entails several steps. An engineer designs an object, for instance, a mouse trap. Once the design is complete, a draftsman draws up a set of blueprints to build the trap. The blueprints are distributed to the manufacturing personnel who build the trap and to the technical publications group who document the use and repair of the trap. Any changes which have to be made during the manufacturing process are documented through the Engineering Change Procedure. The changes are also sent to Technical Publications. Both the original blueprints and the change documents are used to create the artwork required for the user's guide. In the meantime, a technical writer prepares the text that will instruct the buyer on the use of his mousetrap. Once the artwork is sketched and verified for correctness, it is inked to produce legible lines (inking) and photographed to reduce it. Callouts, text which labels and explains the parts of the illustration, are typed up and placed on the artwork (pasteup). The completed manual text is given to the illustrator who inserts the drawings at the appropriate locations. The next step is to photograph the complete page to produce a negative for the production run. The negatives are then used to print the manual.

1.5 TECHNICAL PUBLICATIONS SYSTEM BENEFITS AND LIMITATIONS

All publishing areas agree that the bottlenecks are associated with the pasteup and dummy phases; at best they are an "imprecise art". These cause the page makeup cycle to be the most labor intensive portion of publishing and the most expensive. Previously, phototypesetters allowed the embedding fo commands to determine the layout of the page without a preview station. Later systems included a small CRT for viewing just prior to printing. This was a critical improvement with the introduction of software to merge graphics with text on these systems. One enhancement led to another and large publishing groups, especially those heavily involved in technical manuals, wanted a totally integrated system whereby the illustrator designed, layed out the manual page and proofed it, all on the same video display terminal.

A graphic artist is visually oriented; the creative process is enhanced by viewing the immediate results of a design change. The interactive system also decreases the number of trial and error cycles required to produce a correct copy as well as many other benefits.

> "The increased capabilities of these new workstations are bringing traditional production activities back toward the author and designer, and therefore bringing the source of creativity closer to the tools of the trade. Design and content are beginning to merge into an ongoing creative process. In the long run, these developments should mean not only faster turnaround and lower costs but also increased creative freedom." (5)

The interactive capability allows the artist to archive a drawing which will remain legible and correct fifteen years later; it can be retrieved tomorrow, easily modified and reprinted quickly with minimum effort. Problems exist in the form of user interfaces, inability to handle many art types such as half-tones and color photographs, the lack of a true rendering on the CRT of type fonts available on the phototypesetter and the inability to "walk copy" (to look at several pages of a manual at a time to confirm the smooth flow of thought). The inability of many typesetters to handle half-tone and color art (1) is not a consideration in the area of technical publications since the art is of the line art genre.

Other limitations include the poor user interface and the cost, a complete system usually runs between $95,000 to $160,000 per workstation. The biggest hurdle appears to be the inability to store completed pages in final format for reference and interactive updating of a page unit. The most important, and perhaps the most difficult, requirement is whether a system has the ability to perform both pagination and page updates.

According to the historic data on these systems, improvements are being made. The first interactive graphics workstation was introduced to do page composition ten years ago and today over 2,000 workstations are being utilized (1). One driving force has been the increased complexity of both drawings and manual layouts, rendering the embedding of commands in the text for locating art inefficient. Lead times are becoming shorter while qualified artists who understand the required formats are becoming more scarce. Lastly, our society is becoming more visually oriented as evidenced by the U.S. Government's new manual formats. In the past, the design terminals have had more capabilities than the phototypesetters, i.e., a more complete page could be displayed on the CRT than could be output by the typesetters. These were corrected by the cut and paste method. Other deficiencies include:

- o rotation of types
- o background/foreground reversals (black/white interchange)
- o fancy borders, unusual shapes
- o overlays
- o silhouettes
- o screened half-tones (1)

With the exception of screened half-tones, these problems have been corrected in the new design systems. Perhaps the last factor to be dealt with is cost. The lower cost systems have a large number of limitations but as the cost of electronics continues to decrease, the cost of the system will decrease, making them more attractive. With the present economic situation, even large producers of manuals are having a difficult time justifying such a costly system for one application area.

2. USER REQUIREMENTS

2.1 SYSTEM INADEQUACIES

Many hours of research have produced mechanically viable systems. However, the user interface and drawing capabilities remain adolescent. CAD/CAM systems customized for technical publications often retain old, engineering terminology with foreign input modes. Technical illustrators and artists draw with a pen in their hand which is moving around on a piece of paper. The technical publications systems have a number of different hardware configurations, none of which exactly duplicate the artist's known environment. For instance, all systems have the display unit mounted perpendicular to the floor. The person using the system does not have to look down at a drawing board.

However, an entirely new set of hand-to-eye coordinations must be learned since the hand is moving on a surface removed from the one being viewed. Another problem area is the input mode. One system uses cross-haris which are seen on the drawing screen; another uses a mouse on a digitizing tablet; while still another uses a penlike stylus and digitizer. Users who were surveyed voted for the method they were first taught as the one they preferred. In an attempt to decrease the "culture shock" in going from manual to automated methods, there appears to be less resistance to those systems which most closely resemble the current environment, i.e., a system which uses a light pen to input directly onto the CRT.

Although these are pertinent issues for the illustrator's comfort, a learning process must also occur to draw on the computer system. The artist must choose some type of drawing action which is done by selecting a menu item with the input device; light pen, crosshairs, trackball, finger, etc. The menu may be located on the screen, on a floating digitizing surface, or on a menu panel underneath the screen. The menu selection requires the operator to take some form of action. For example, drawing a circle may require selecting the circle function from the menu, indicating the center on the screen with the input device, and keying in the radius of the circle. The artist has just used two hands to push buttons, key in numbers and located a point on the drawing surface. To further complicate matters, this sequence often involves menus, keyboards and/or drawing surfaces which are out of the direct viewing line of the CRT, i.e., a new hand-to-eye coordination pattern.

Another problem arises from the use of keyboards. Most artists are unfamiliar with typing which makes systems requiring large amounts of keying inhibitive. Frustration arises from the "hunt and peck" method, thereby decreasing the system's effectiveness.

2.2 USER REQUIREMENTS

The computer industry has spent a great deal of time on the ergonomics of a system, CRTs swivel, tilt and, if mounted on a pedestal, raise and lower. European standards go so far as to specify the dimensions of the keyboard. Some companies allow the user to dispense with the keyboard by making it part of a menu. Another system feature is the ability to support multiple menus or a combination of a menu and a set of function buttons to supply additional capabilities. These are nice but are often located in awkward positions. Figure 2 presents the "ideal" workstation as conceived by the author and several illustrators.

Unfortunately, the amount of time spent designing the hardware configuration has not always been duplicated in the user interface and system capabilities. The illustrators interviewed were not as concerned with whether the CRT could be viewed in bright light or whether it had extensive data management functions as they were with the actual drawing fuctions. Software features which rate highly are:

(1) an interface to existing word processors —allows typists to input majority of text,
(2) a simple, relevant user interface—instructions in meaningful terms,
(3) the ability to merge graphical and textual data on the same plot—allow the placement of text from a word processor on the drawing,
(4) direct isometric input—most systems support isometrics by creating the orthographic first,
(5) rapid hidden line removal, preferably automatically,
(6) easy rotation of drawings for different viewing perspectives,
(7) multiple viewing screens with a universal update feature,
(8) standard library of parts,
(9) ability to transfer drawings from existing CAD systems used by Engineering to the technical publications system,
(10) electrical schematics options such as dragging symbols to a given horizontal position.

2.3 SUMMARY

Several factors have been discussed which pertain to technical publications drawing requirements. These include such features as making the keyboard a menu, interfacing to the in-house word processor, hidden line removal, rotation of drawings and standard library of parts. Another factor is the ability to transfer drawings from existing CAD/CAM systems to the Technical Publications system.

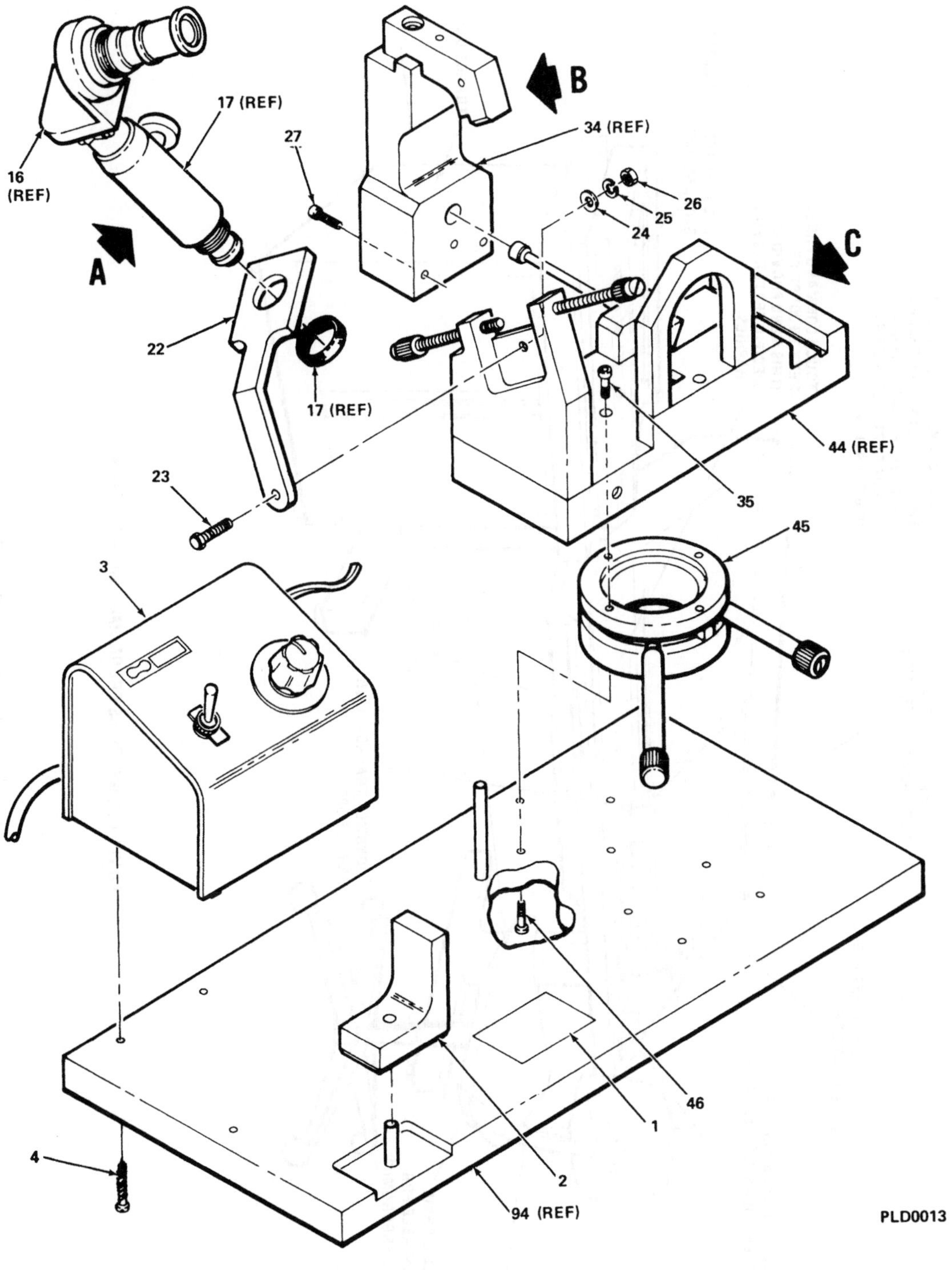

Figure 1: Typical Illustration Used in a Technical Manual

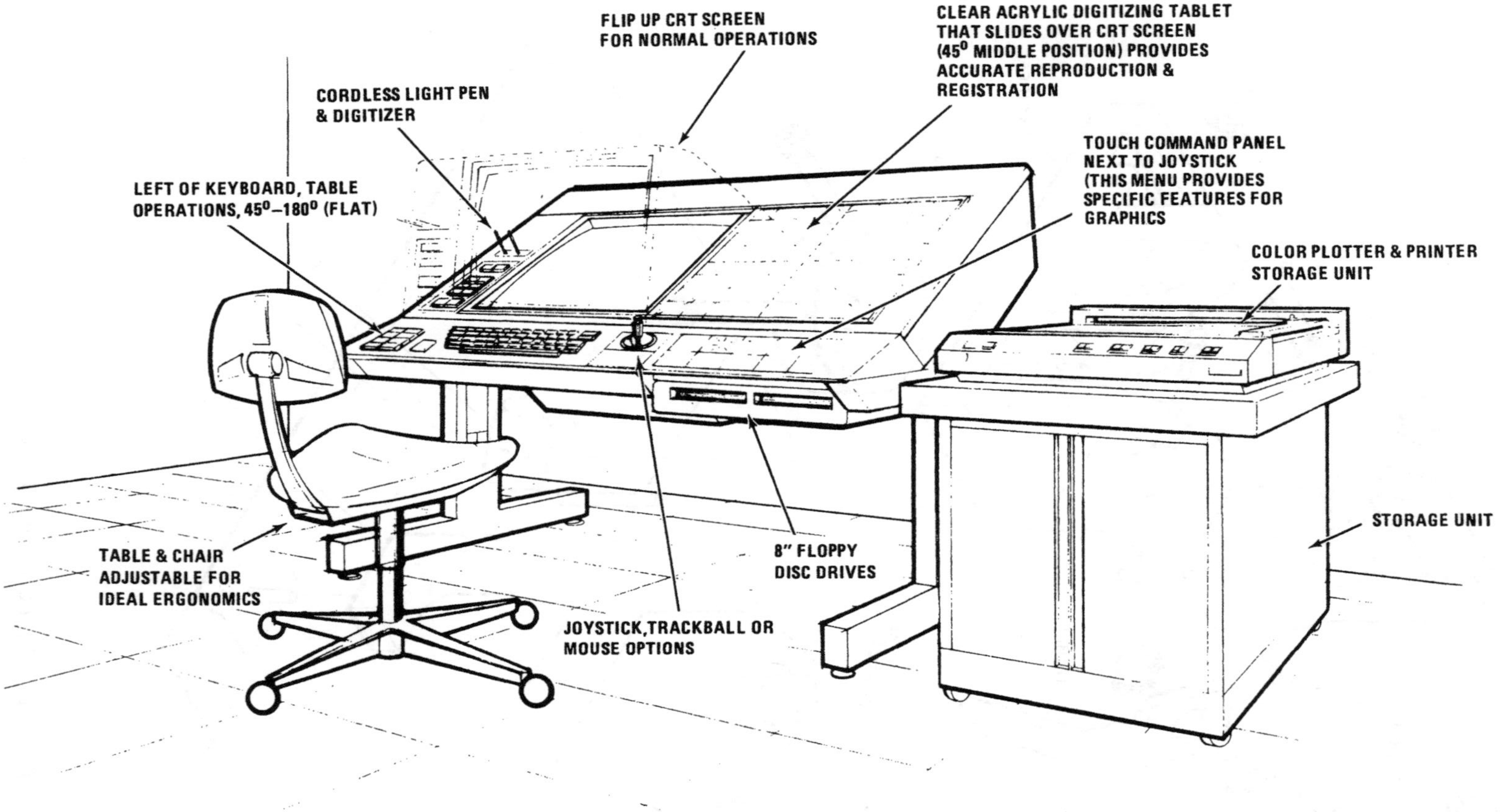

COMPUTER GRAPHICS ERGONOMICS WORKSTATION

Figure 2: The Artist's Workstation

Much concern has been directed toward the user interface and rightfully so. Management has been or will be faced with retraining their employees to use automated methods. Resistance to learning new thought processes and drawing techniques is prevalent, especially among older artists. Any tool which will make the transition easier is appreciated.

3. THE PROPOSED TECHNICAL PUBLICATIONS SYSTEM

3.1 IDEAL SYSTEM

The ideal system (Figure 3) is one which takes the design from birth to death. In this ideal system, a new product is conceived in the Product Design area, goes through several design cycles, and is then released to the Engineering department. This department runs stress, vibration and thermal tests. Once these constraints are satisfied, the designs are sent to the draftsmen for the blueprints to construct the product, and to Technical Publications for user manuals. At the same time, Manufacturing checks the feasibility of construction techniques before the design is finalized. The Numeric Control (N/C) personnel use the geometric data stored in the model to create and verify tooling paths prior to entering into the manufacturing stage. In the future, the robotics instructions will also be generated at this stage. Problems encountered during any phase can be resolved before the design and blueprints are finalized, thereby preventing costly mistakes during the manufacturing process.

It is also easier to locate and modify a drawing on the automated system. This system will speed up the entire process since the draftsmen, manufacturing engineers and N/C programmers will have access to the data for early previewing. Technical Publications may also access the product and begin work on the manuals as time becomes available. Any changes can be easily sent to each department via the computer network, achieving a more dynamic design and update process. The production time will become shorter, allowing manufacturers to respond to consumer needs faster. This network would provide the stepping stones necessary to eliminate the shipping of tons of maintenance/user manuals. With the ability to communicate between all systems, technical manuals could be transported from the producer's system to the user's system via a telecommunication link or magnetic tape(s). Manual updates would be made quickly and efficiently through the telecommunication link, decreasing documentation costs. A fully automated system, as described, would eliminate paper as the major exchange medium; although laser printers and old, paper-oriented work habits will continue to propagate paper manuals for some time.

3.2 SYSTEM CONFIGURATION

A fully distributed, networked system creates the ideal system—workstations double as standalone units for special assignments and as automatic CPU backups. Even with the decreasing costs of electronics, one workstation per CPU remains expensive and impractical. The ideal hardware configuration is a CPU which supports two or three graphics stations to decrease the host cost. Admittedly, graphic applications require an enormous amount of computing power to create and maintain the picture.

Many functions are moved into the workstation with results sent to the CPU at regular intervals to update the disk copy. Local processing includes shading of objects, color selections, dynamic rotation, movement, scaling and clipping. Drawing functions for lines, circles, arcs, ellipses and surface definition are handled at the terminal. High volume functions like hidden line removal, splining, curve fitting and updating drawing files are performed by the CPU. The true representation of the phototypesetter fonts requires a large portion of the graphical database and should be done at the CPU. An outline representation should be available for initial construction and layout of pages to accelerate the process. Central functions performed at the CPU range from library management to communicating with other users.

Many other tasks which are compute intensive should be placed on a devoted workstation (CPU) to prevent response degradation. One such example is the translation of graphic data into drawing instructions for the phototypesetter. The lone workstation, the central workhorse, services auxiliary disks, handles tape drives and a printer, translates outside

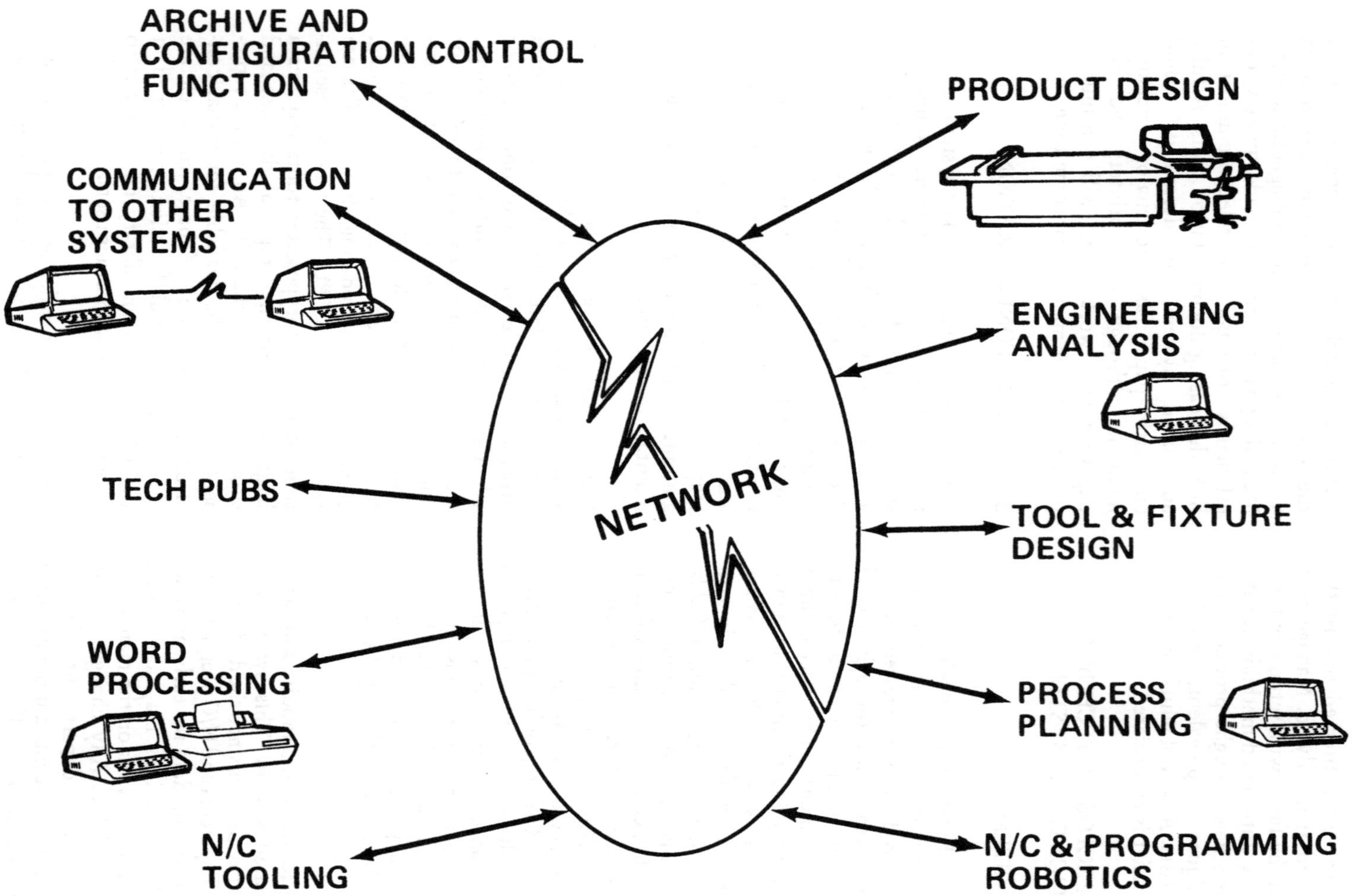

Figure 3: Ideal Design System

word processing data into graphical data to be used on the drawings, drives the high resolution plotter and phototypesetter, collects operating errors from the network and acts as an auxiliary text input station. The workhorse is responsible for activating backups of each workstation through the system timer. The entire network is booted from here. Most likely, the workhorse will be located in a computer service center where a trained operator monitors the system. This alleviates training each artist/operator how to care for and feed the workstation/computer. To the user, an unknown being automatically backs up individual disks and handles other maintenance details. Managers and supervisors need not worry about these details; they continue to manage their illustrators who now have a new tool.

Since the technical publications group is now part of an even larger network, they can request/receive specific instructions from a designer when the drawings are requested. A message is sent across the network to the designer when the need arises. If the person is not on the system at that time, the message will be placed in a wait mode. This will decrease the amount of time spent looking for a person and, perhaps, the phones will not ring as often.

Another advantage is the availability of drawings. They can be previewed for time estimates prior to actually receiving the job. Drawings are grouped together on the system for easy job identification and not lost under a stack of miscellaneous designs on someone's desk.

An illustration is a composite of several engineering drawings which often become separated on the checker's table. By using an intelligent naming scheme, it is possible to group them together in the computer. Engineering changes are more dynamic since the change is sent universally and immediately through the system. Design problems can be averted before costly construction and documentation are done. For instance, Tooling is asked to preview a preliminary design by Engineering and finds that a bolt has been placed where there is not enough clearance. The mechanic cannot disassemble the part without first removing three other items. Engineering is requested to move the bolt during the first design review. Manufacturing does not create an impractical part nor has the documentation begun. Engineering does not continue with the design that would have been changed after it had gone through analysis, drafting of blueprints and checking. The company has saved money and time because the mistake was caught early and the "Change Procedure" was not activated. Employee morale is higher since the amount of rework is decreased.

3.3 USER INTERFACE

The illustrator is seated at an ergonmically designed workstation, the world at his fingertips. Blueprints can be accessed from any draftsman to create illustrations or modify a design. The drawing's owner can be queried on a particular area. The drawing can be modified to create the right perspective and labeled with text created on a word processor elsewhere. A manual page can be created and sent to be printed on the phototypesetter. The system capability exists to perform a virtually unlimited number of tasks but the user does not know how. This is a person who communicates via telephone, a drawing pencil and paper. The illustrator receives several blueprints with the expectation of creating a complete part. Next, descriptors and text are pasted on the drawing to form a complete manual page. The page is then photographed to create a master negative for production runs. Management is now asking this same illustrator to pick up an electronic device to point with and to type instructions into a black box to produce a drawing or to talk to Engineer Smith three blocks away. Is Management crazy? How does all that "stuff" increase productivity? One such example is Figure 4. The original plans required nineteen blueprints, eight of which were changes. The changes are often insertions and/or erasures which are quickly accomplished on a computer system.

Tools must be provided to make the new situation as much like the old, familiar one as possible while adding the many innovations. This will decrease the amount of resistance to using the system as well as the training time.

For instance, several systems provide a freehand sketching capability which is the artist's normal activity. The system, however, must

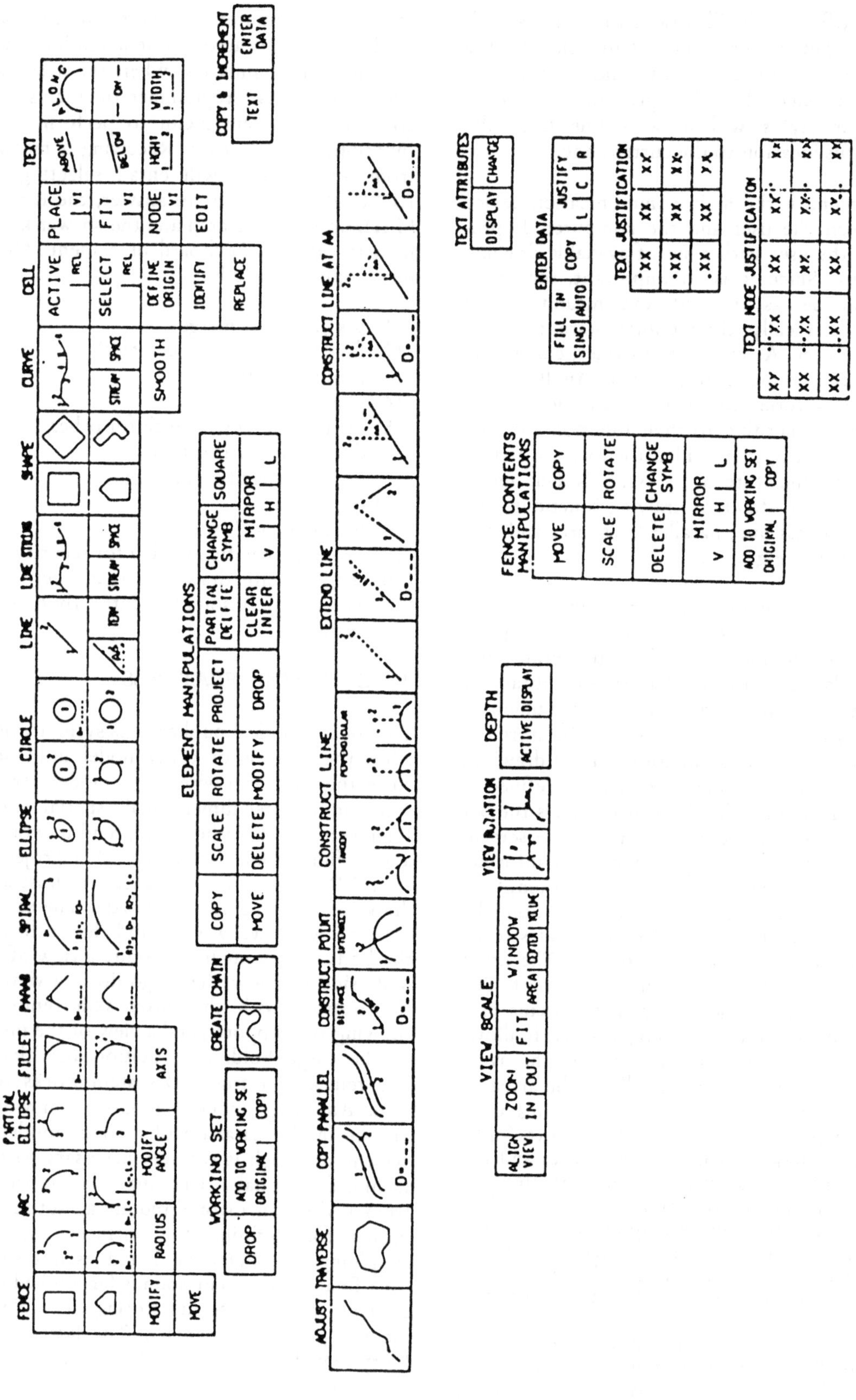

Figure 5: Sample Menu

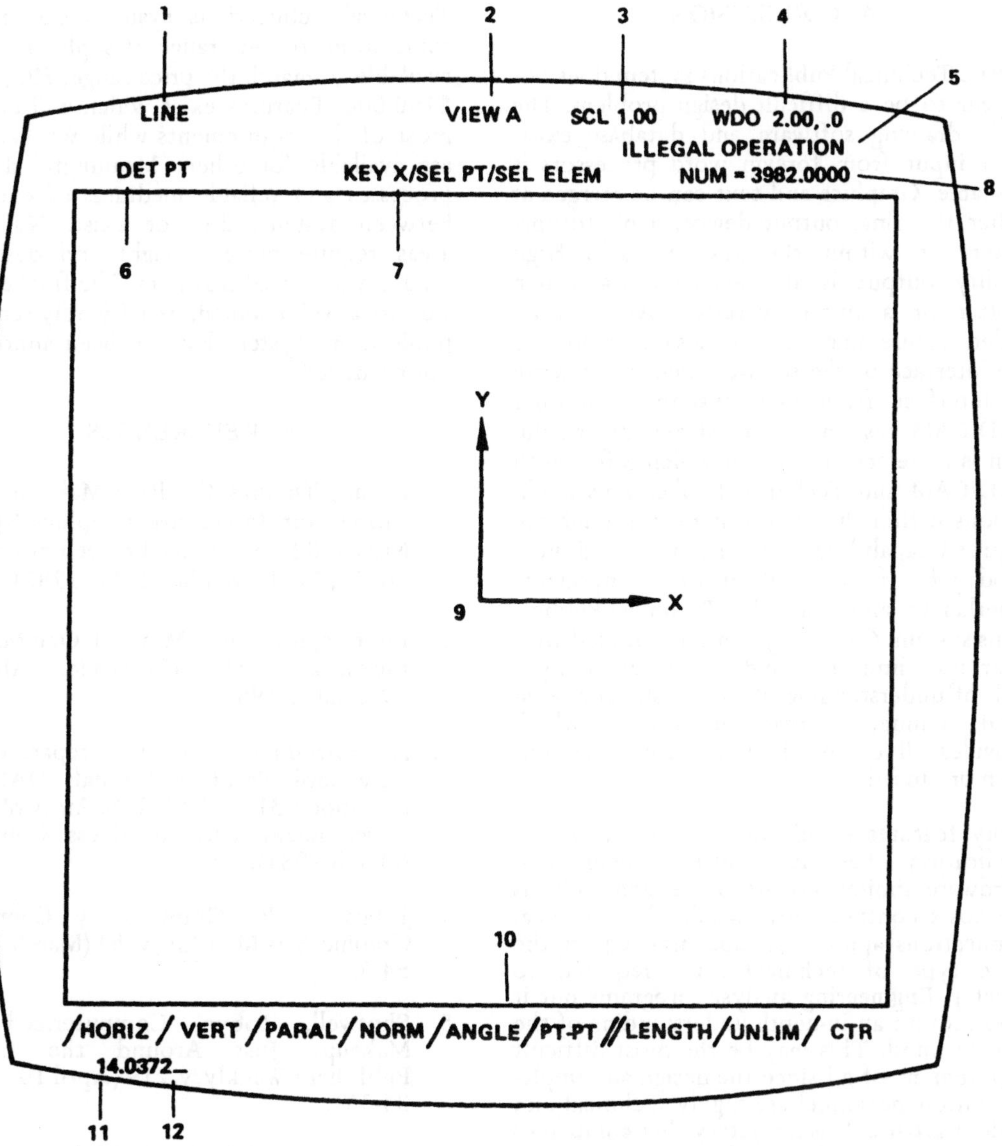

1. FUNCTION
2. VIEW IDENTIFIER
3. SCALE
4. WINDOW PARAMETERS
5. ERROR INDICATION
6. ACKNOWLEDGEMENTS
7. COMPUTER MESSAGE
8. COUNTER
9. ORIGIN INDICATOR
10. MENU
11. TEMPORARY HOLD REGISTER
12. NONDESTRUCTIVE CURSOR

Figure 6: Example of Menu Prompts Viewable on the Display Screen

4. CONCLUSION

A true Technical Publications system does not appear to be a difficult design problem. The basic drawing software and database exist. Text input from foreign word processors is available. Graphics and text can be merged at either the final output device, a phototypesetter, or within the system itself. High quality output is also available via a pen plotter or a phototypesetter. And, as an added feature, many companies either provide the interface or the software needed to write an interface from their system to another CAD/CAM system. The most expedient solution is to select one vendor which offers both CAD/CAM and Technical Publications applications within the same system. Although the required capabilities exist, no one has done a good job of tying all of the components together to supply a viable Technical Publications system. Current systems are riddled with poor user interfaces and are governed by a lack of understanding of the artistic and technical manual process. A system which provides all of the desired features has not been produced yet.

Many features should quickly appear as the application becomes better understood. Hardware limitations are decreasing daily as the costs continue downward. The technical publications application does not require the same type of technical skills required to develop Engineering analysis programs but it does require an in-depth understanding of the artistic mind. This may be the most difficult requirement of all since the design and implementation personnel are highly technical and seldom artistic. It is a mystery that companies developing technical publications systems have not hired several experts to advise them. Surely, CAD/CAM and AEC (Architectural Engineering Construction) systems were not developed without consulting with the experts in these areas. The investigator considers this but more evidence of the immaturity of the application. When questioned about the industry's late arrival in this area, one salesman cited lack of funds for purchasing a technical publications system. The technical publications group is considered an overhead cost and is not an essential part of the production process. If this is true, where would we be without manuals to instruct us on the operation and maintenance of cars, airplanes, electrical appliances, garage door openers, etc.?

Technical publications systems do exist in some form. A wide range of sophistication is available to match the price range, \$95,000 to \$160,000. Features exist which will perform most of the requirements while workarounds are available for others. Meaningful drawing processes and reliable mediums of exchange between systems do not exist. Numerous areas require more thought and design to render a truly useful system. The final conclusion is: a well planned, user friendly technical publications system has not been announced nor produced.

5. REFERENCES

1. Hagan, Thomas G., Page Makeup Applications for Interactive Graphics System. NCGA '82 Conference Proceedings, vol. II, no. 21, 22 (November 8, 11, 1982), 1.

2. Intergraph Product Material, One Madison Industrial Park, Huntsville, Alabama (December 1982)

3. An Introduction to the Preparation of New Look Technical Manuals. DARCOM Handbook 310-1.1-81, U.S. Army Materiel Development and Readiness Command (March 1981).

4. Johnston, P., Graphics by Computer. Graphic Arts Monthly, v. 54 (March 1982), 54-6.

5. Shotwell, Robyn, Computerized Page Makeup: Just Around the Corner. Publisher's Weekly, v. 219 (April 10, 1981), 34-36+.

6. BIOGRAPHY

Deborah G. Brown is a member of the Data and Communications Systems at Ford Aerospace and Communications. She is currently involved in developing interactive computer graphics systems for engineering analysis functions. She has also developed a graphics training program for the user community. Previous experiences have been in the areas of CAD/CAM dealing with both 2D and 3D systems.

Brown holds a B.S. degree in mathematics from Stetson University and an M.S. in computer science from the University of California, Irvine.

10

IMAGE PROCESSING TECHNOLOGY APPLIED TO ENGINEERING DOCUMENT MANAGEMENT SYSTEMS

Rebecca D. Costa
OMEX
2323 Owen Street
Santa Clara, California 95051 U.S.A.

Abstract

Recent progress in image processing technology will impact on those intending to acquire engineering documentation management in the near future. The climate in Washington towards image processing technology has recently taken an abrupt turn away from scepticism mixed with caution, towards the release of serious RFPs. Representing one of the suppliers of image processing systems, a number of technical requirements specified in many government RFPs have been reviewed. Described are some of the key ingredients specified by those intending to automate the process of storing, retrieving, and electronically distributing engineering documents. A realistic view is presented on which of these ingredients are available and which require further development on the part of both government and commercial industry.

One unfortunate aspect of systems technology, whether we are talking about data processing, CAD/CAM, image processing, or any other computer system, is that historically all parts have not evolved equally. Every system is a unique combination of input, output, storage, and communications technologies, some state-of-the-art and others still in the infancy of development.

In my opinion, it is critical to differentiate on a component by component basis, between those technologies which satisfy real user requirements and those which require further enhancements and development. By examining the ability of today's technologies to satisfy the requirements of a complete engineering documentation, storage and retrieval system, you, as potential installers of such equipment, can determine far in advance which requirements can be met today and which are better met by tomorrow's technology. Further, you have the opportunity to design RFPs based more on existing componentry than on development and, to this extent, you will be able to minimize risks associated with installing a new technology.

I must say, regrettably, that not all of the technologies needed to satisfy today's complex engineering records requirements have evolved equally. More specifically, I must point out that the input and output peripherals needed to accommodate today's diverse engineering records have lagged behind the development of storage, communications, database and configuration management technologies. But before I describe in detail the input/output peripherals needed, I would like to review the types of engineering records being discussed, as well as examine the evolution of methods of storing engineering documentation.

Long before the advent of CAD/CAM, even prior to the micrographics revolution, when drawings were hand drawn with nothing more than paper, a wooden drafting board, a pencil, and the patience of a genuine craftsman, drawings were archived in their original form on paper.

Well, as any of us who have owned a paperback novel or an old photograph for more than ten years knows, paper has qualities which do not lend themselves to preservation for prolonged periods of time. Within a relatively short period of time, the integrity of images on paper begins to deteriorate as the paper itself degenerates, becomes brittle and difficult to handle.

About the time paper records were becoming cumbersome and difficult to manage, our friends in the film industry came along with a process which enabled records managers to capture complex paper images on a small piece of 35mm film. And film, claiming to retain image integrity behond 100 years, became almost impossible to resist. The storage economies, manageability, and long life of film seduced an entire engineering industry into purchasing many millions of dollars worth of camera and development equipment quickly, enabling paper drawings to be converted into roll film and fiche records.

Sometime later, about the time engineers began getting used to manual film readers and the idea that drawing originals really could be retained on a small piece of 35mm film, Holorith-encoded aperture cards were introduced to engineering records managers across the country. These cards permit drawing numbers, classifications, revision numbers and other important document identifiers to be key-punched into computer cards which also contain the filmed 35mm image. Overnight, aperture card sorters enabled engineers to locate any single or group of drawings in record-breaking time, reducing resources needed to manually sort through many, many pieces of film.

Aperture cards represent a desirable mix of technologies: the ability to capture the storage and archival benefits of film, while simultaneously taking advantage of a computer's number-crunching and accounting capabilities. As a result of this combination, the greatest percentage of engineering records today are stored in aperture card format.

But we face another transition, a transition from hand-drawn and filmed records to a technology used to manufacture engineering records in a format not compatible with film or aperture card technology. That technology is CAD/CAM.

As those familiar with CAD/CAM know, CAD/CAM produces graphic images while simultaneously producing non-graphic engineering data for stress, heat, and vibration analysis, bills of materials, and so on. CAD/CAM generated engineering drawings are a combination of three-dimensional graphic imagery and invaluable nongraphic analysis data. As a result, separate from drawings created on CAD/CAM systems, analysis data must also be preserved. The filming process is only useful insofar as it can capture a static image. It is not capable of preserving any intelligent data relating to a CAD/CAM design. Therefore, archiving CAD generated records on 35mm film results in nothing short of throwing nongraphic information out with the bathwater.

So, many engineering records managers find themselves in a bit of a dilemma. A small percentage of original records remain on paper, a small but increasing percentage of CAD/CAM drawings are being preserved in digital form on magnetic tape, and, for the most part, engineering records are still being maintained in non-computer readable format on film and aperture cards.

But the future is clear; if there is one thing agreed upon, it is that the world is going digital. The use of CAD/CAM systems is escalating at 30 to 40 percent annually and, as the cost of design stations drops, we can expect this escalation to continue. The number of records which must be archived in direct digital form is on the rise, while the number of records which are adequately preserved by film are on the decline.

So far, I have described how records have evolved to encompass three very diverse media forms: paper, aperture cards, and archival digital storage. I have also suggested that not all the technologies needed to satisfy the requirements for a complete records management system are available today. I will now address the requirements for a system which can store, retrieve, distribute, and manage paper, aperture cards, and digital engineering records.

The requirements for any engineering documentation system can be broken down into six basic components: INPUT, which includes the ability to accept data from such devices as paper, micrographics, CAD/CAM and magnetic media; OUTPUT, which refers to printers, plotters, alphanumeric and image displays, CAD/CAM and Computer-Output Microfilm (COM) peripherals; STORAGE, which includes magnetic disk and tape, on-line macrographic storage and optical mass storage; COMMUNICATIONS, in terms of both local area and remote networks; and finally, CONFIGURATION and DATABASE MANAGEMENT, which provide the ability to rapidly access and manipulate engineering records.

To describe an engineering documentation system, let us begin with input requirements. As pointed out earlier, engineering records are currently stored on three basic media: paper, aperture cards, and magnetic tape. With this in mind, it is important for a complete engineering documentation system to have some mechanism for digitizing records on film and paper and converting them to electronic form for rapid distribution to digital displays.

In other words, a flat bed scanner capable of accepting original drawings stored on paper is needed. Second, a device capable of quickly reading Holorith-encoded data, selecting aperture cards and digitizing batches of filmed images is also needed. And third, not least of all, an engineering records management system must be capable of accepting graphic and nongraphic input directly from CAD/CAM systems.

These three input technologies represent the cornerstone of any complete engineering documentation system. But this is only the beginning; what goes in must come out. So what are the output requirements of an engineering documentation system?

Output is generally divided into two categories: softcopy, referring to the ability to view information on a display terminal; and hardcopy, the ability to produce a printed version of information on paper or film. A successful engineering documentation system must be able to accommodate a wide range of hard and soft copy output devices. For example, many requests for archived engineering records may be satisfied by simply providing the ability to view the drawing at an unintelligent display terminal. Workers on a manufacturing floor or in a service department may only require the capability to periodically view a part number or an assembly.

But imagine the detail lost if we reduce a full E size drawing to the size of a normal display; the resolution would hardly be sufficient for readability. Therefore, inexpensive display terminals must provide the capability to view an entire drawing, while also incorporating the ability to select and explode a specific area of a drawing. Pan and zoom capabilities are a requirement for display terminals associated with engineering documentation systems.

Ideally, both documentation originally stored on film as well as the digital form whould be viewable on the same terminals. But today. images on film are converted into raster data, while CAD/CAM usually creates images as vector information. As a result, some sort of raster to vector conversion process is required for images stored on film to be displayed on CAD terminals. Conversely, for vector images to be viewable on unintelligent image displays, vector data must be converted into raster information.

While intelligent CAD workstations and unintelligent image display terminals may meet much of the need to quickly reference an archived engineering record, the requirement for hardcopy output on paper and microfilm will never be completely eliminated. The reason? For one thing, it is often easier to work with paper which can be marked up and red-lined than to work off a soft display. And it is still far less expensive to run a quick plat of a drawing which must be used for many days than it is to tie up an image terminal continuously for the same period of time.

Also, microfilm is and will continue to be a convenient and sensible method for assembling vendor bid sets. Microfilm is inexpensive, can be easily mailed, and requires only that vendors involved in bidding a contract own an inexpensive manual reading device. Although I have no doubt that one day all prospective government vendors will own compatible image processing systems so that engineering records associated with bid packages can be electronically transmitted to appropriate vendors, until that day, bid packages probably will continue to be distributed in film form.

detail. Because the solid model is complete, each section is complete and requires no manual completion or editing. Because the information comes from an engineering solid modeling database, the views and sections produced are dimensionally accurate. This allows using the automatic dimensioning capabilities of the CAD/CAM system with no need to manually input dimension values.

8. SOLID ASSEMBLIES

In addition to producing single parts, the solid model can be used to produce entire assemblies, simply by merging the solid models of individual parts. One of the difficulties in producing assembly drawings using traditional techniques is that the detailed part drawings usually do not contain the view necessary for maximum clarity in the assembly, the view that best shows how the parts fit together.

Solid modeling can easily produce any view or even a set of views. Parts can be added or removed from the assembly as required. Assembly drawings with hidden lines removed can be produced.

Exploded assemblies can be produced simply by moving the individual solid parts away from each other.

It is possible to take section cuts through an assembly, with complete control over which parts will be sectioned. This sectioning can be used to reveal inner parts and to add clarity to the assembly.

9. ASSEMBLY SEQUENCES

A common problem in manufacturing is to clearly show a sequence of operations necessary to produce a finished product. Traditional engineering drawings are often difficult to interpret on the shop floor, leading to mistakes. A series of photographs showing the operations to follow is good, but at least one part has to be made to obtain these photographs.

Solid modeling can produce the needed sequence of pictures, either as hidden line removed wireframes or as color-coded shaded images.

An example of this that has occurred in production involved a complex weldment. The weldment contained over twenty pieces, and used a series of jigs and fixtures during production. Problems were encountered with pieces being welded out of sequence, so that it was impossible to install all of the pieces, pieces being welded out of place, and jigs being welded inside the part.

A series of solid models were made, and shaded images produced. These models showed each step of the assembly. The pictures were color-coded, with the existing (old) pieces shown in one color, the new pieces shown in another color, the jigs to be removed shown in another color, and the jigs to be used to hold the new piece in yet another color. Also, the base that supported the entire weldment during manufacturing was shown in still another color. Simply by looking at the picture, the welder could tell which piece to add, where it went, and which jigs to add and which to remove for each step.

These assembly sequences can be made from the engineering solid models, and can be made long before metal is cut for the first part.

10. TECHNICAL DOCUMENTATION

Beyond engineering documentation is an area often referred to as technical documentation. Technical documentation is a combination of text and illustrations. Examples of technical documentation include user's manuals, service and repair manuals, process planning instructions, trouble shooting guides, parts catalogs, and even advertising. All of these documents require accurate illustrations of how the part or assembly will appear to someone who is trying to either use or repair the product.

One of the most difficult problems is a part catalog. Any type of complex product has to have catalogs so that parts and part numbers can be determined when it becomes necessary to replace them. The illustrations for parts catalogs require accuracy and often need exploded assembly drawings. These catalogs also use very large numbers of illustrations. All of this adds up to large investments in time and manpower to produce a catalog. These catalogs can be produced much more efficiently with the aid of solid modeling.

IGES
Version 3.0
(In Work)

- Solids Modeling Representation
- Further Electrical Extensions
- Further Finite Element Modeling Extensions
- Plant Design Extensions
- Manufacturing Technology Extensions
- Drafting Enhancements

McAuto

GP31-1728-15

FIGURE 31

Existing Alternates to IGES

- Standard Data Format (SDF)
- Standard Interchange Format (SIF)
- Section 5 of ANSI Standard Y14.26M (ANSI 5)
- Direct Translators

McAuto

GP31-1728-10

FIGURE 32

SDF
Standard Product Data Exchange File Format

- **Existing Translators (Vought Developed)**
  - — CADAM
  - — CADDS-3
  - — PATRAN
- **Translators Under Development**
  - — Prime
  - — CADDS-4

McAuto

GP31-1728-9

FIGURE 33

SDF
Standard Product Data Exchange File Format

- **Developed by Vought Corporation**
- **Key Element - Efficient Data Packaging**
  - — System and Application Independent
  - — Compact
  - — Packed Binary File
  - — Extendible
  - — Minimization of Associativity Pointers

McAuto

GP31-1728-8

FIGURE 34

SIF
Standard Interchange Format

- **Developed by Intergraph Corporation - 1979**
- **Key Elements**
  - Ease of Implementation
  - Readability
  - Ease of Editing
  - Flexible and Compact Format
  - ASCII and Binary Forms
  - No Pointers
  - Free Format/Keyword Driven
- **Existing Translators**
  - Applicon
  - Autotrol
  - Calma

McAuto

GP31-1728-7

FIGURE 35

ANSI Y14.26M Standard
Section 5
(ANSI 5)

- Geometry Represented as Relational Structures
- Rigorous Definitions
- Topology Tied to Underlying Geometry
- Minimal Number of Structures
- Extendible
- Curves, Surfaces, and Solids Represented

McAuto

GP31-1728-40

FIGURE 36

International Standards Organization

- **Organizing New Technical Committee**
  - — Industrial Automation (Technical Committee T. C. 184)
  - — December 1983 - Paris

McAuto

GP31-1728-11

FIGURE 37

International Standards Organization

- **Anticipated Subcommittees**
  - — Numerical Control Equipment
  - — Numerical Control Programming Languages
  - — Database Exchange
  - — Robotics
- **Probable Submission of ANSI Standard Y14.26M for Consideration as International Standard**

McAuto

GP31-1728-12

FIGURE 38

PDDI

Product Definition Data Interface

Deliverables

Task I

- Test Procedures
- IGES Evaluation

McAuto

GP31-1728-13

FIGURE 39

PDDI

Product Definition Data Interface

Deliverables

Task II

- PDDI Exchange Standard Format
- Working Form of the PDDI Format
- Access Software

McAuto

GP31-1728-14

FIGURE 40

Program Summary

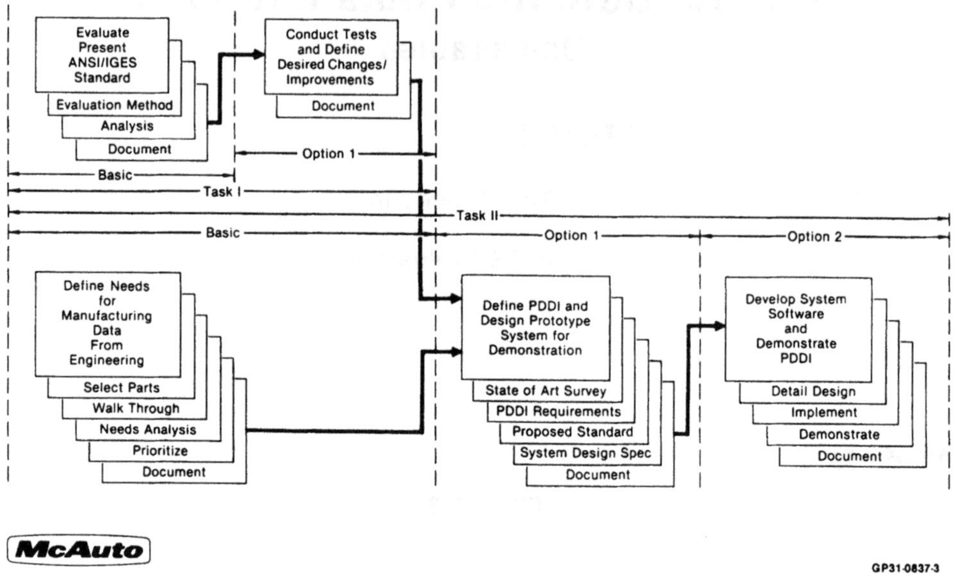

FIGURE 41

ICAM Product Definition Data Interface (PDDI)

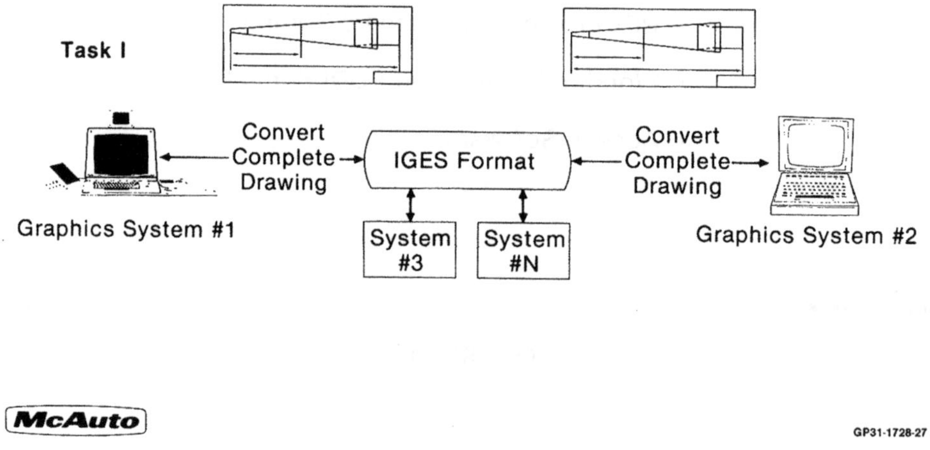

FIGURE 42

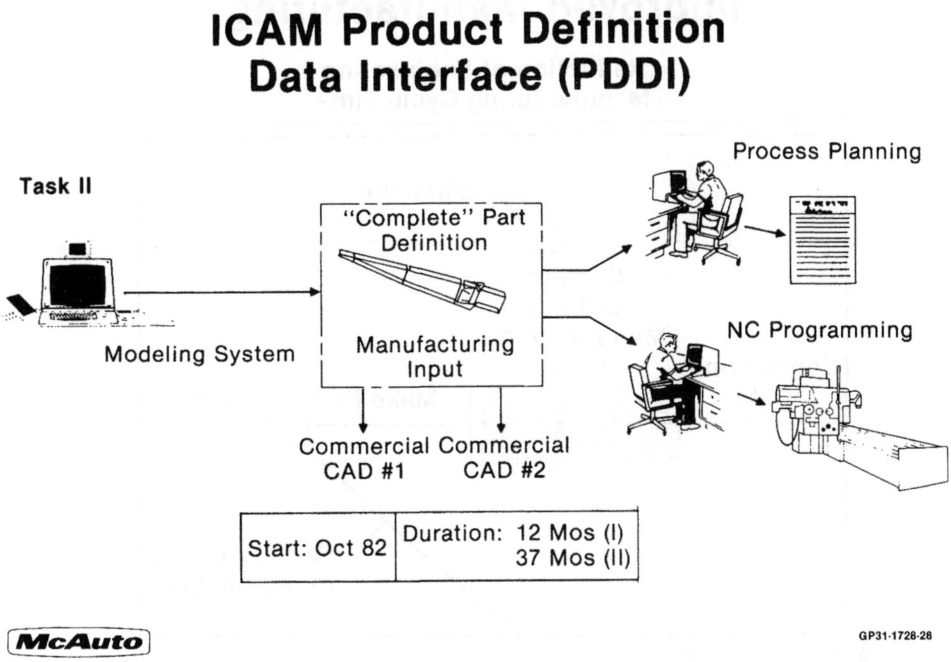

FIGURE 43

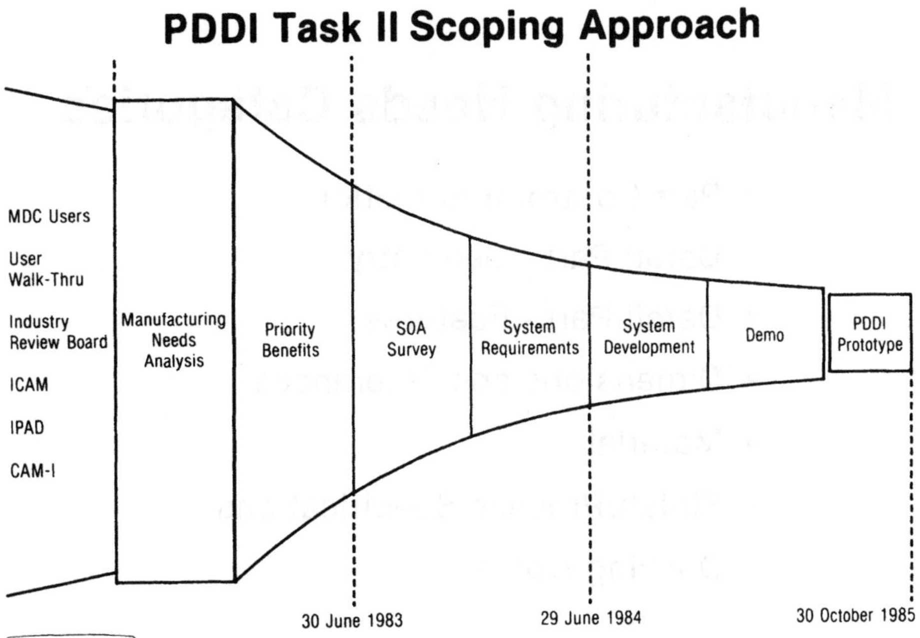

FIGURE 44

Improved Manufacturing

Reduction of Engineering Manufacturing Cycle Time

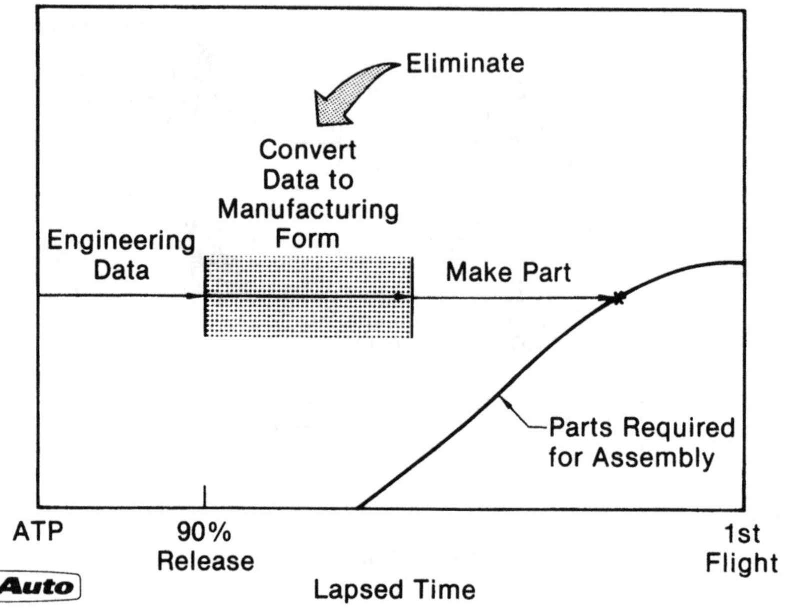

McAuto

GP31-1728-41

FIGURE 45

Manufacturing Needs Categories

- Part Control Information
- Detail Part - Geometry
- Detail Part - Features
- Dimensions and Tolerances
- Material
- Finish/Process Specifications
- Drawing Notes

McAuto

GP31-1728-42

FIGURE 46

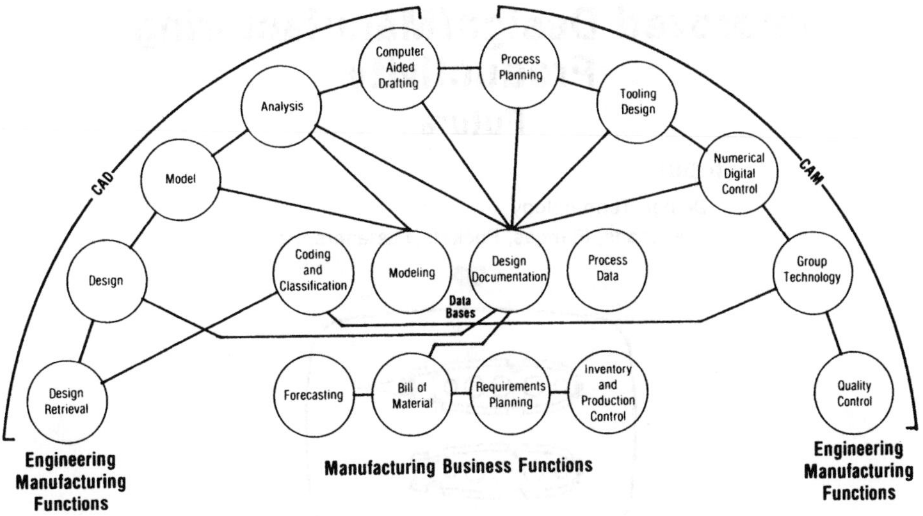

FIGURE 47

Improved Design/Manufacturing Productivity
Present

Input

Drafting Terminology
- Lines, Circles, Curves

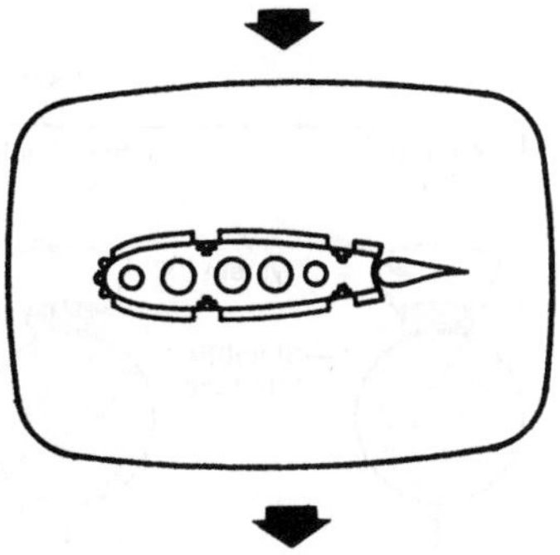

Output

- Manufacturing Interpretation
- (Design/Manufacturing) Productivity Losses

GP31-1728-43

FIGURE 48

Improved Design/Manufacturing Productivity

Future

Input

Design Terminology

- Loads, Flanges, Pockets, Fasteners, Holes

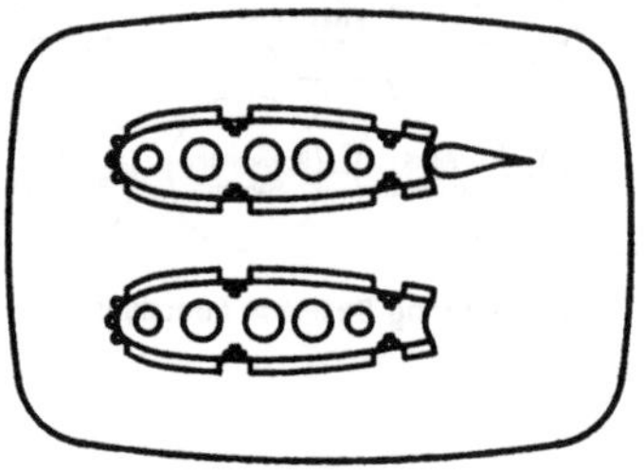

Output

- Increased Design/Manufacturing Productivity
- Broad Operations Usage
- Automation Enabling Product Data Base

McAuto

GP31-1728-44

FIGURE 49

PDDI Deliverable Product and Its Possible Use

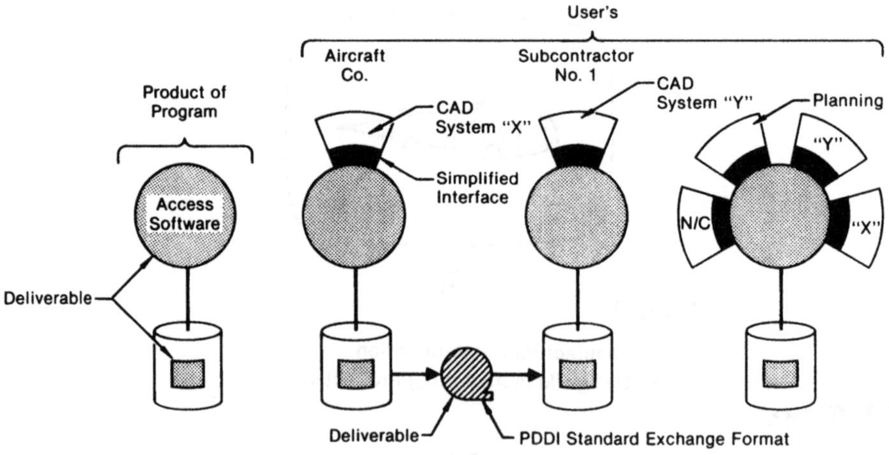

McAuto

GP21-0564-129

FIGURE 50

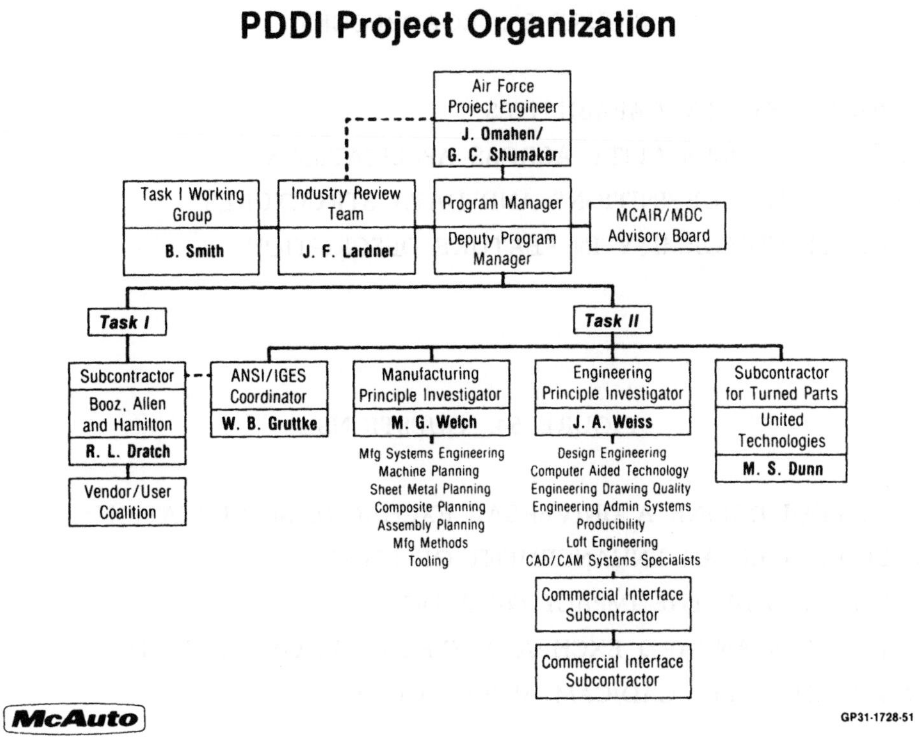

FIGURE 51

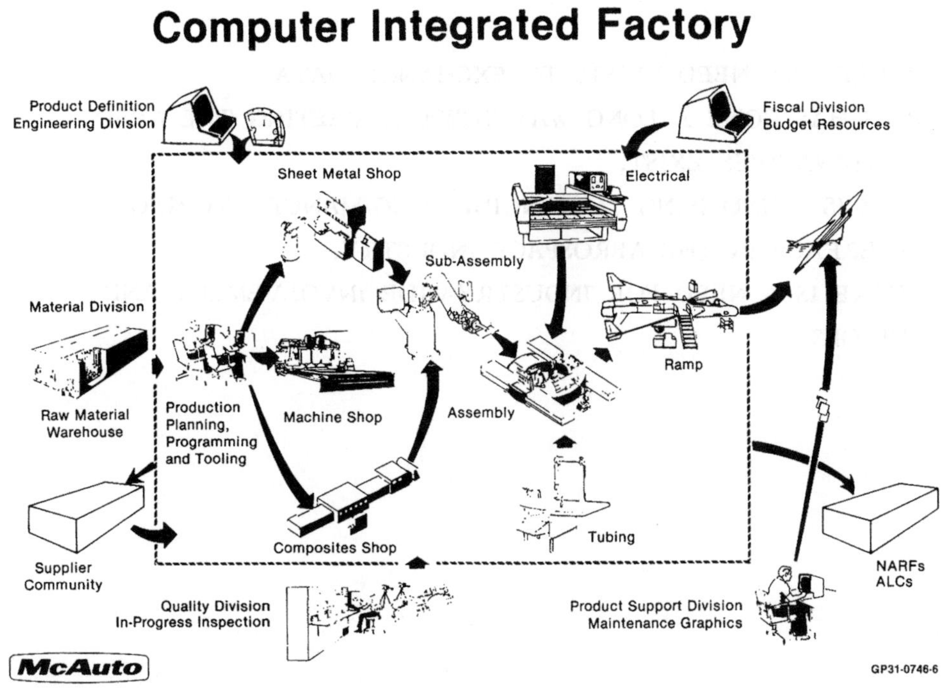

FIGURE 52

FIGURE 53. ROADBLOCKS

- o CURRENT SYSTEM CAPABILITIES
- o LACK OF COMMONALITY ACROSS APPLICATIONS
- o PROPRIETARY DEFINITIONS AND DATA STRUCTURES
- o LACK OF UNIFORMITY IN FEATURE DESCRIPTION

FIGURE 54. QUESTIONS

- o WILL PDDI DEFINE A UNIVERSAL EXCHANGE SPECIFICATION?
- o WILL PDDI BE AN ENHANCEMENT OF IGES?
- o WHAT WILL BE THE IMPACT OF PDDI?
- o CAN ONE STANDARD EXCHANGE SPECIFICATION MEET THE NEEDS OF "ALL" CAD/CAM APPLICATIONS?

FIGURE 55. SUMMARY

- o A DEFINITE NEED EXISTS TO EXCHANGE DATA.
- o IGES HAS GONE A LONG WAY TOWARD MEETING THE NEED.
- o ALTERNATIVES EXIST.
- o PDDI IS ADDRESSING THE ENGINEERING/MANUFACTURING INTERFACE IN THE AEROSPACE INDUSTRY.
- o THERE IS A NEED FOR INDUSTRY-WIDE INVOLVEMENT AND SUPPORT.

11. PRODUCTION OF TECHNICAL DOCUMENTATION TODAY

The preparation of a page for a manual or catalog is usually handled through the process of paste-up. Production of a page starts with two separate processes—text and graphics. Text is usually entered on a word processor, for output on either a printer or a phototypesetter.

The graphics are either prepared manually by a technical artist or prepared on a graphics system and output on a pen plotter. After the graphics have been prepared, they are pasted down on an oversized "master" sheet, which has the page outline marked on it.

There is no text at all on these graphics—neither handlettered text nor text from a pen plotter has high enough quality to achieve the professional appearance that is often called "reproduction quality". Instead, each individual callout, note, or other text on the graphic is output from a phototypesetter, cut out with an "X-acto" knife, and separately placed on the page.

Any corrections or changes to the graphics must be performed by hand. With several different layers on the master, this may be very difficult to do. Any changes to the text require another pass through the phototypesetter, which may require a day or more of turnaround time, and must then be cut out and pasted down.

The body of text for the page may be output as one long page, called a galley, and then cut up and pasted down in blocks.

Alternatively, a computerized page composition system may be used to fit and position the text on the page.

Manual page preparation is labor intensive, time consuming, and error prone. Final quality depends on the skill and care of the person doing the paste-up. In addition, revisions and updates tend to be difficult to perform and manage and may even require starting over at the very beginning.

The final result is a page master, which must be photographically reproduced to make a printing plate. Individual page masters must be carefully stored, so that the manual or catalog can be reproduced when necessary. Large companies may have to store millions of pages of these masters.

12. NEW DEVELOPMENTS IN PRODUCING TECHNICAL DOCUMENTS

A recent innovation in the area of interfacing CAD/CAM and technical documentation is the combination of text and illustrations directly on the system. Until recently the primary output device of CAD/CAM systems has been the line plotter—perhaps a Benson pen or electrostatic plotter.

Phototypesetters have been used in the publication industry for a number of years because of their combination of quality and versatility. The text in books and magazines which has what might be called a professional appearance is typeset text. Typesetters are able to produce exceptionally high quality text. Digital CRT phototypesetters are computer controlled, and can have hundreds of thousands of fonts in their software. They can produce text in a wide range of sizes, from a few millimeters to several centimeters. They are able to produce proportionally spaced text (an "i" has less width than an "m") or mono-spaced text (typewriter style). Typeset text is both more readable and more compact than typewriter style text. Typeset documents will usually require one-third less space than the identical documents produced on a typewriter or letter-quality printer.

Until recently, phototypesetters have been limited to text only.

The restriction of phototypesetters to text has been largely due to the way people have used the machines rather than any inherent hardware limitation. Recently developed software packages allow output of illustrations, in fact outputting the combination of illustrations and typeset text, in a single pass on a phototypesetter. Interfaces between the CAD/CAM graphics system and either an Autologic APS-5 phototypesetter or a Xerox 9700 laser printer allow direct output from the CAD/CAM system to these devices, utilizing the full range of text fonts available on these text oriented machines. This has several advantages. It allows electronic editing of both the text and graphics, and produces a dramatic improvement in the time required to generate completed pages.

13. CAD/CAM AND TECHNICAL DOCUMENTATION

The CAD/CAM system is used to produce illustrations and can also be used for interactive page composition. In addition to interactive text entry, some systems have the ability to read in text from an external ASCII text file—such as the file produced by a word processor or text editor—and add this text directly to the drawing file.

These capabilities allow preparation of an illustration using all of the graphic construction and editing features available on the CAD/CAM system, particularly solid modeling. Text for notes, callouts, part numbers, etc. is interactively added and positioned on the page. At any point in this process, the text or illustrations can be edited, deleted, or other text or illustrations merged in.

If interactive page composition is being performed, body text and illustrations can be merged at any time. The illustrations can be scaled (changed in size) to fit in available space. The text can either change size with the illustration or stay a fixed size. Different illustrations on a single page can be scaled independently.

One of the primary advantages of interactive composition is the ability to see what the final results will be before typesetting. This combination of previewing and editing greatly increases the probability of getting the desired output on the first try.

Quality is an important byproduct of using a CAD/CAM system in technical documentation. A computer model is very precise. Horizontal and vertical lines are exactly horizontal and vertical. Text is exactly parallel to the bottom of the page. Lines are a precise weight; patterned lines (dashed, center, phantom, etc.) have exactly even spacing at all points; crosshatching is absolutely spaced. A CAD/CAM system allows producing consistent, high quality work faster than can be done by hand.

The fact that pages and artwork exist as computer files rather than as physical masters provides several advantages, in addition to electronic editing. These files can be stored on-line, so that many different people can access or copy them. If the CAD/CAM system is part of a computer network, any user on the network can access the files. If security is a concern and only certain people are to be allowed access, the computer's file protection facilities can be used. Archival storage makes use of magnetic tape or disk packs that are more compact than physical page masters. Multiple copies of the files can be made and stored separately to reduce the possibility of loss or destruction of the files.

The ability of computers and CAD/CAM systems to store, retrieve, archive, copy, list, and in general manage large numbers of files simplifies the job of document control.

The ability to generate an entire manual or catalog electronically even opens up the possibility of demand printing using a laser printing system such as the Xerox 9700. The ability to use the computer to update only changed pages will allow keeping documents fully up to date with minimal effort. Data can also be transmitted through a computer network for demand printing at a remote site.

In addition to output on paper, it is possible to generate an entire manual on microfilm, using the microfilm capability of an Autologic APS-5 or equivalent machine.

14. SOLID MODELS AND TECHNICAL DOCUMENTATION

What are the advantages of producing solid models for documentation?

With a CAD/CAM system, the solid model created in the engineering department as part of the design process can be directly picked up and used by the technical illustration department.

Solid models are far superior to simple wire frame models because they allow the illustrator to generate a complete hidden line removed view or set of views at will, simply by changing viewing parameters. Assemblies can be produced by simply merging several individual part models, and exploded assemblies are produced by using the editing features of the CAD/CAM system to move the separated parts. Once again, any needed hidden line removed views can be generated at will for the assembly.

An integrated solid modeling based approach helps the technical illustrator by allowing direct use of the solid models, regardless of where they are produced; by providing shaded images or hidden line removed drawings; and by allowing full use of the CAD/CAM system to "touch up" or enhance the illustrations.

15. CONCLUSION

Solid modeling, in conjunction with a CAD/CAM graphics system, can significantly reduce the amount of time required to produce engineering drawings, such as detail drawings, assembly drawings, or exploded assembly drawings.

The shaded images produced by Solids Modeling (tm) enhance the ability to visualize a part or product and are valuable for their ability to easily convey information. Typeset text and typeset graphics, combined with the interactive page composition capabilities of the CAD/CAM system, can greatly reduce the time required for production of technical documentation containing illustrations or other artwork—especially if models, particularly solid models, are already available on the system.

Over the next several years, CAD/CAM systems and solid modeling will have as much impact on engineering and technical documentation as they have already had on design, analysis and manufacturing.

16. BIOGRAPHY

Russell J. Doty is the Manager of Market Analysis at Applicon. He joined Applicon in early 1980, and has been active in the development and promotion of numerous products at Applicon, including Solids Modeling, Solids Modeling II, and the Technical Documentation. He received a B.S.M.E. from the University of Oklahoma, and has done graduate work at the University of Missouri.

Previous experience includes work for McDonnell Aircraft as a programmer on their CAD/CAM system, and extensive use of the system for mechanical design.

This paper was presented at the November 1983 ATI Conference and is an updated version of an earlier paper presented at an NCGA meeting, copyright 1982 by the National Computer Graphics Association, and reprinted with permission of NCGA.

12

THE IMPACT OF OPTICAL STORAGE TECHNOLOGY ON IMAGE PROCESSING SYSTEMS

Gerald T. Durbin
Storage Technology Corporation
14008 Maebeth Drive SE
Huntsville, AL 35803 U.S.A.

Abstract

The recent announcement of commercially available high density optical storage devices will have a profound impact on the information processing industry. Just as the initial introduction of random access storage created entirely new processing strategies, optical technology will allow dramatic changes in the storage, retrieval, and dissemination of engineering drawings and other pictorial or text-based documents. Storage Technology Corporation has assumed a leading role in this arena with the introduction of the 7600 Optical Storage Subsystem, and the formation of its Systems Integration Division, chartered to incorporate this new technology into deliverable total systems. This paper explores the impact of optical storage technology from the perspective of a leading-edge manufacturer and integrator.

1. EVOLUTION

Until now, processing of high-volume text and image data has been constrained by the cost-effectiveness of on-line storage. The industry has seen rotating magnetic storage evolve along two increasingly distinct paths—performance and capacity. Performance devices are characterized by relatively low capacity and fast access time, and are essential for key input-output functions inherent in today's large-scale response-oriented systems. Capacity devices, on the other hand, sacrifice access time due to the higher number of stored bytes per data path. These new devices store as much as 2.5 GBytes of data on two nonremovable spindles. As these devices approach the practical limit of thin-film magnetic technology, the required tolerances will dictate continued use of non-removable media.

For the high-volume data user, system design involves a combination of performance and high capacity direct access storage, with magnetic tape as the principal backup/archiving medium. However, even with the latest high-capacity disk and high performance magnetic tape, the user must trade off on-line availability against cost of storage, and must live within the constraints of elaborate archival and restore mechanisms which results in access times measured in minutes or hours. The application must still be constrained by the limits of the technology.

Ideally, the solution to this problem would embody the random access capabilities of magnetic disk (at a reasonable level of performance), as well as the removability, transportability, and archivability of magnetic tape. In addition, one would hope to surmount some of the inherent problems of magnetic tape as an archival medium, such as erasability and short shelf life, while providing access within seconds to vast amounts of data at an affordable price.

Recent announcements of digitally encoded optical storage systems, such as the Storage Technology 7600 Optical Storage Subsystem, promise a solution to this problem. The STC 7600 offers numerous advantages which meet the requirements of imaging applications:

(1) Capacity: The STC 7440 Optical Media Unit (platter) contains 4 GBytes of user data on one side of a 14" disk. Through attachment to an automated library, hundreds, or even thousands of these platters could be available on-line within seconds.

(2) Performance: The 7640 Optical Storage Unit provides an average access time of 85 msec, with data transfer rates up to 3 MBytes per second.

(3) Archivability: The media provides a life of 10 years without loss of data integrity, with a bit error rate of 1 in 10^{13} (or one bit in error every 312 platters). In addition, the non-erasable nature of the recording media reduces backup requirements dramatically.

(4) Availability: Since the media is removable, it can be made available on-line very quickly. In addition, it lends itself to transport easily, without concern for damage from temperature, x–rays, or magnetic fields.

(5) Affordability: The cost of an optical solution is less than that of combined magnetic disk and tape solutions while eliminating the data volume constraints of those technologies.

2. APPLICATION

To understand the diversity of the image processing industry, one need only compare an electrical circuit diagram with a page from a standard insurance policy. Each represents a major segment of the burgeoning image processing market. The advent of optical storage is reshaping the cost-justification formulae for on-line storage. Documents which would never have been considered for computerization are now likely candidates for data capture. When the growth of office automation is combined with the availability of optical technology, the concept of a paperless office filing system moves much closer to reality.

The successful implementors of these new systems will have three major distinguishing characteristics:

(1) Manufacturing-level expertise in both magnetic and optical storage devices.

(2) A dedicated staff of systems integrators, with sufficient software expertise to ensure a working large scale system incorporating multi-vendor hardware.

(3) A world-wide maintenance and support organization.

In order to give the reader a better understanding of the next generation of image systems, the next three sections discuss system capabilities and characteristics from the perspective of a leading manufacturer and integrator. Three areas will be discussed: Input, Storage Control, and Output.

3. INPUT

While many techniques exist for image input, the two most commonly used are laser scanning and Charged Coupled Device (CCD) scanning. At present, laser scanning technology has a number of deficiencies which preclude cost-effective scanning of documents or engineering drawings. CCD scanning, on the other hand, has improved greatly, and now allows high speed/high density scanning of virtually any type of input, from microfilm to 18-foot-wide hard copy. This most recent CCD technology provides the foundation upon which to build a new generation of intelligent scanning workstations.

The scanning workstation is a critical component of any image processing system. While it is unrealistic to expect any single workstation design to be universally applicable, a "standard" scanning workstation might have the following characteristics:

(1) Each workstation would be a self-contained module capable of scanning, editing, and storing data. Output from the workstation might be transmitted to a central computer, or simply stored on a local device.

(2) The scan rate would be operator-selectable for 100 to 800 points per inch. This is an important technological breakthrough which will allow the scan to be adjusted to the document or drawing being processed.

(3) The scanner will be designed for 11" x 17", two-sided capture. This will allow the front and back of forms to be processed in the same pass, and multi-page forms to be handled much more efficiently than previous technology would allow. The potential exists to quadruple the throughput of existing systems through two-sided scanning of multipage forms.

(4) The captured image would be immediately available for operator review, and will undergo on-line adaptive processing, compression, character recognition, indexing, or reformatting under local control.

(5) Facilities will be available for hard copy, microfiche, and aperture card scanning. The data stream will be modifiable within the workstation to accommodate the various types and sizes of media.

In summary, the newly developed scanning technology will be integrated into powerful, microprocessor-based workstations. These workstations will function in both standalone and communicating mode to provide an efficient and flexible entry point for image data.

4. STORAGE AND CONTROL

Once image data has been captured, it is part of the information system, whether resident in an individual workstation or a central archive. The system design must provide control of the data through its life cycle of storage, transfer, and dissemination. An integrated system will make use of several different types of storage:

(1) High Performance Magnetic Disk – Relatively low capacity devices with extremely fast access times. These devices are necessary to store system programs and indices necessary for high-volume multi-user environments. In many cases, a solid state device such as the STC 4305 may be used in this role.

(2) High Capacity Magnetic Disk – High capacity devices such as the STC 8650, or the STC 8380, which stores 2.5 GBytes of data on two spindles of non-removable media. These devices provide the primary random-access memory for the system, and would be used to hold active data.

(3) High Density Magnetic Tape – Used for conventional sequential storage and periodic backup of magnetic disk memories. The STC 4670 tape subsystem is an example of this technology.

(4) Optical Storage Units – The STC 7440 will store 4 GBytes of data on one side of a 14-inch removable platter. The platters are provided in easily-loaded cartridges for protection during handling. The recording is permanent (non-magnetic) and each platter holds the equivalent of 40 reels of 6250-bpi magnetic tape, with a bit error rate of 1 in 10^{13} over a ten-year shelf life.

In addition, optical platters could be stored in automated libraries for very high volume users. These optical data libraries would vary in size from a few hundred to over a thousand cartridges, with access to the requested cartridge in seconds without operator intervention.

This approach creates an archive previously unattainable, with capacities in the 1-5 TByte range. Libraries will support multiple optical drives, so that the subsystem can be configured with greater flexibility.

The optical platter is a key element in the overall system design. Its 4 GByte capacity and compact size allow it to be used for data transfer as well as archival storage. Platters produced at standalone workstations can be transported to a central system where they can be loaded either manually or through an automated library. Users will also have an interesting new mode of high speed data transmission. Consider the following: assuming a high speed 50KB telecommunications line between workstation and central system, it would take over seven days to transmit the contents of a 4 GByte platter. Alternatively, the user could simply send the platter via overnight mail, effectively transmitting all the data in a matter of hours. The fact that the optical media is impervious to magnetic interference makes this strategy even safer than shipping today's magnetic tape.

One of the most promising features of the new design is the ability to store both intelligent (computer processable) and non-intelligent (imaging) data on the same media. This capability results from the all-digital nature of the system, and signals a new era of processing flexibility. The workstation operator will be able to process both numeric and non-numeric data, and conversions within the system may be used to achieve an optimal mix of intelligent and non-intelligent data for a given application.

The storage and control of data in future systems will encompass a variety of hardware devices, all of which must function under or be compatible with a single set of software. This raises two key issues: first, the integration of the equipment must be technically feasible; second, someone must create integrated working systems. As a leading manufacturer, STC has responded to each of these challenges.

In order to insure software compatibility among all devices, STC has standardized the optical storage unit on the IBM MVS SP 1.3 Operating System. In this configuration, the optical device will communicate through an STC 8880 Controller which attaches directly to an IBM-type channel. Since all of the other devices discussed are also supported by IBM software, the result is complete operating system compatibility. In addition, an Intelligent Peripheral Interface (IPI) has been announced which will allow the optical storage unit to operate in a non-IBM environment.

The second challenge required an even stronger response. As the need to incorporate optical technology into complete functional systems became clear, an entire division was formed. The Systems Integration Division has been chartered to integrate a variety of STC and non-STC products into functional total systems which address specific user requirements. Major components such as workstations will be standardized and used as the building blocks for customized user systems.

5. OUTPUT

This section discusses in general terms the complex problem of data distribution. The utility of any information system is largely a function of its ability to disseminate data in a timely and functionally useful way. In image processing systems, the problem is complicated by the high volume of data and the variety of media. With respect to volume transmission, the optical platter itself provides an innovative vehicle, as described in the previous section. For actual output, however, a variety of devices will be used to satisfy unique user requirements. Given the all-digital nature of the system, there are many existing devices well suited for integration into image processing systems.

Hard copy output can be produced using several current techniques. Engineering drawings and documents up to "B" size may be reproduced by either laser or raster printers, and those above "B" size by raster printers. The engineering drawing is restricted in quality by the capability of the printer and its ability to image on sensitized paper. Laser printers offer high speed production capabilities and high quality images. These printers are capable of printing up to 120 pages per minute with excellent quality. Both raster and byte data may be transferred from the digital storage system directly to the laser printers for reproduction of the information onto hardcopy.

Digital data may also be routed to Computer Output Microfilm (COM) equipment for generation of high resolution images on film. In document systems, this information may be generated onto standard NMA formatted microfiche or roll film. In the engineering drawing environment, the information is transferred first to indexed silver film and then placed in special film-to-card diazo duplicators that operate under control of the central computer system to output the requested number of duplicate aperture cards. It is also well within the realm of reason to consider the replacement of existing archival COM with the optical disk itself. Since the information normally written to COM is from a digital data base and the COM output process converts the digital to analog, there are no provisions to reverse this process should the data be requested again. The optical disk will allow the incorporation of the digital data into an all-digital information system without the penalties that currently exist with analog microfilm devices.

A third distribution system consists of visual displays. The display workstation is comprised of a microprocessor-controlled display with the ability to process both intelligent and non-intelligent information on the same screen. Data from local area networks, or from remote transmission, is stored and displayed under complete local control. The local workstation would also have its own disk storage and printing capabilities. Since this is a vital link in the information flow process, each display station would be a workstation in itself, fully capable of interaction with the host computer or standalone operation as a remote microprocessor for additional cost effectiveness.

The combination of hard-copy, optical archiving, COM, and visual display will provide output capability to meet a variety of user needs. All of this hardware is currently available and is being integrated into innovative system solutions.

6. CONCLUSION

The age of high technology image processing systems is here. Recent technological advances in this area now provide the tools that will allow introduction of all-digital technology into an arena that has long suffered from the constraints of analog-based design.

In addition to providing a critical technological breakthrough in the introduction of optical storage, Storage Technology Corporation has embarked on a concentrated effort to offer serious users a solution to their information processing problems. The incorporation of non-intelligent information with intelligent data from computers now allows the capture and dissemination of information that previously has been unavailable. The key to successful implementation will be the integration of newly available scanning and storage technology into customer systems which utilize standard hardware modules and are backed by a world-wide support organization. Optical storage has opened the door to a new era in image processing.

7. BIOGRAPHY

Gerald T. Durbin is currently affiliated with Storage Technology Corporation. Prior to this he specialized in the marketing of large data base systems and he has been instrumental in the design and installation of numerous real-time computer systems and engineering documentation systems. Mr. Durbin received his B.S. degree from Indiana University.

13

MASS STORAGE SYSTEMS FROM THE USER POINT OF VIEW

Jack K. Gingrich
ALPHAREL, Incorporated
Westlake Village, California 91361 U.S.A.

Abstract

Much has been said about optical storage disk and mass storage systems by various suppliers. However, very little has been said about these systems from the user's point of view. This paper discusses what one user thinks the "super system" should look like: what the required inputs, the operation, and the actual retrieval mechanisms are. It discusses integrating the system to current operating procedures and the impact on manufacturing, engineering, and marketing. The risks and concerns from the user standpoint are also discussed. This paper deals with an engineering drawing data base but it can easily be applied to any mass document storage system.

1. INTRODUCTION

About a year ago, optical disk storage systems were discussed only at various technical meetings such as this Automation Technology Institute Conference. Most people felt that it would probably be around someday—usually defined as three to five years away. But now just about everyone seems to have heard of the optical disk; all information-related societies are now emphasizing this new technology and it's hard to pick up a technical magazine that does not have some information on the optical storage disk.

Much has been said about the development of the media but relatively little about the user. Suppliers coming on the market are advertising the advantages of their systems. This may or may not correspond to what the user really needs. I am going to discuss the subject of mass data base storage systems from the user's point of view. I have been a user for over 25 years and, as Manager of Engineering Services, have been continually besieged by people from Engineering, Manufacturing, Marketing, and everywhere else wanting faster and more reliable information.

So, this is what I think a system should be—sort of a super system. I will use engineering drawings in my description because I have been deeply involved with this massive data base. However, the system could easily be changed to apply to any mass document storage system.

An engineering file can contain millions of documents, tracings, aperture cards, and in some cases even microfiche. In General Electric Medium Steam Turbine, there are over one million aperture cards in six large rotating drawer files. It takes five people full time to pull the cards, run copies, and refile them. And there are the normal problems of cards out-of-file, misfiled, or mutilated. A like number of tracings exist and space is rapidly running out. Tracings exist in files, on top of files, in boxes, literally everywhere. It requires three people to handle these tracings and there are always instances of lost and/or badly mutilated tracings.

It had long been realized that this was an antiquated technology and falling way behind the technological advances of other functions. The need for an automated system was well understood, but which one? There are some automated systems which use microform as the base. But advanced technology points strongly to a digital storage system. So the "super system" I shall describe deals entirely with a digital system.

2. INPUT ASPECTS

Starting a digital system is a very difficult and time-consuming task. The documents must be changed from hard copy to a digital format, properly identified, indexed, and stored on an optical disk in logical sequence. An efficient loading system to the optical disk pays dividends later during retrieval. The following are the input methods:

2.1 APERTURE CARD SCANNER (also a microfiche scanner in some cases)

By allowing say eight seconds to scan each card, it would take 56 weeks, eight hours a day, five days a week, to load one million cards. Obviously an automatic feeder to the scanner is required and it would be a good idea to have several scanners during the start up to speed initial loading.

A quality control program is necessary to assure that the scanning will be accurate. After the system has passed its acceptance test, it will probably be necessary to scan only one out of every five cards, and then as confidence builds up, graduate up to one out of 25, 100, 200. In order to handle the QC efficiently as the cards are scanned, they are stored on a magnetic disk buffer. Periodically, maybe once a day (depending on the number of scanned cards stored), the system manager will direct that the magnetic disk data be transferred to an optical storage disk. It is conceivable that even this QC activity could be automated by periodically placing a target aperture card in the system and comparing that target electronically with the target's pattern stored in the system. As technology improves, this could become a reality. But for now, the option is to inspect the cards visually.

2.2 HARD COPY SCANNERS

Even in the perfect system, I am convinced that drawings will continue to be drawn on drafting boards in spite of the great advances made by IAG and automated design. Therefore, a hard copy scanner will be required. Actually, two scanners are necessary in this super system.

The first scanner will handle only "A" size documents and, to make things perfect, will include an automatic sheet feeder. The second scanner will handle "B", "C", and "E" sizes. The "A" size document should be scanned in ten seconds or less, these documents should be the same as outlined for the microfilm cards.

2.3 IAG DATA

Most companies have some sort of IAG equipment; therefore the super system will accept data directly from the IAG system and allow it to be retrieved, all automatically and completely transparent to the IAG operator. The way things are handled now, the data is transferred from the IAG working disk to tape or to a magnetic disk and then later transferred to tape. To revise a drawing that exists on tape, an operator must take the tape file and transfer it to the working disk and then, when the revision is complete, reverse the procedure, which takes time. Tapes are not archival and must be rewound and rewritten periodically which is also a bother. Storing all this data on optical disk eliminates this problem.

2.4 OTHER SOURCE DATA

The automated design which is now rapidly catching on results in other large data bases that exist on various main frame or mini computers. These can tend to overload a computer, or at least result in substantial storage cost. It will be necessary for the super system to accept and store this digital data and also allow for retrieval as required.

3. OPERATIONAL ASPECTS

Now that the data has been scanned and stored, we will discuss how to handle it. Note that the data comes from many sources, including different computers. Data will flow into the storage media from IAG in vector format, from scanning in raster format, from sources such as IGES in neutral format, and from word processing in character format.

There are those who claim that data from IAG should be stored in neutral form. If this storage is intended to be archival (10+ years), it is quite possible that the protocol by the IAG supplier will be changed so that the stored protocol would not be recognized by the new protocol and therefore could not be retrieved. If the data were stored in neutral format, it would always be retrievable. This has some merit; however it has disadvantages. Converting a file to neutral format and back to vector is very time consuming and a neutral file contains a lot more bits and thus is more cumbersome to handle. It seems more logical to store the data in the original vector format and, should an IAG supplier eventually change the protocol, ample notice will be given—requiring a couple of weekends of work to change the old file to the new protocol file.

One other consideration is important for the vector format data. Vector format cannot be easily transferred to a hard copy format except by fairly slow drum and flat bed plotters. To use the faster electrostatic and laser printers, a raster format is required. Converting vector to raster takes time—too much time for a dynamic system. Therefore, the system should convert the vector data to raster data. Both vector and raster representations of the drawing will be stored. Someone wanting a copy goes to the raster format and an IAG operator goes to the vector format. It is all automatic as far as the user is concerned. Because the storage media is inexpensive, the duplicate storage will not be a problem.

A unique feature which ALPHAREL and INTEGRATED AUTOMATION are claiming is the ability to change a drawing which is stored in the raster format. This change capability opens a whole new world in handling a drawing storage system. A raster scanned drawing is called up on a CRT (with change capability) by a drafter, within 20 seconds maximum. He can then make actual changes right on the screen. The operator can delete graphics or text, and add circles, arcs, lines, and text from word processing if appropriate, and thus have a new revised drawing in a fraction of the time it would have taken to do it manually. When the revised drawing is refiled, it will be filed under the next revision (i.e., Rev. A to Rev. B). When a stored drawing is requested from the file, the latest revision is always retrieved. The system automatically picks out the latest revision. By special arrangement, an authorized person may request an earlier revision. Quite often it is necessary to go back to an earlier revision on a drawing; for example, a turbine component that has failed and needs to be rebuilt was built to Rev. C but the latest drawing is now Rev. R. Drafters have spent a lot of time working a drawing back from Rev. R to Rev. C. With this system, it can be retrieved immediately. Incidentally, this is an advantage of the optical disk quirk that data once stored cannot be erased.

If you can make changes to a drawing in the digital state, the question arises, why keep a tracing at all? Drawings originated on the IAG system are kept only in vector format, no hard copies other than the aperture card. I have already discussed the problem of stored tracings. One good way to solve this is to get rid of them and use an all digital drawing data base. Of course, this can give us some nervous moments—we must make sure that we don't lose our data base. We will have to write everything in duplicate—either as it is written or later as the disk is filled. At this time, I prefer the latter procedure.

Another excellent feature of this digital storage system is the ability to enhance the image at the time the document is scanned. We have some very old prints in our file which are torn, ragged, with ink and pencil on the same drawings, and extraneous data such as coffee stains. In one example, we sent an "E" size drawing to Japan and our customer sent it back saying that they couldn't read it. They were right, we couldn't read it either. The only way we could have corrected it was to redraw the entire drawing—estimated at 80 hours. Unfortunately we did not have the time to do it, so our customer redrew it—very embarassing to us. If we had had the line enhancement capabilities of the scanner, we could have saved ourselves a lot of trouble. This enhancement capability in itself will save a lot of time and money.

Because we are talking about storing over one million drawings, it means we will need in excess of 100 disks, assuming a one gigabyte disk. To store all of these disks on a single drive system would be impractical—100 drives would be cost and space inefficient. Obviously a juke box arrangement is required. As a user, I would like to see a juke box in the 50 to 100 disk category. There are some juke boxes under design holding as many as 256 disks. This appears to be too many as one disk drive is too little. I prefer to hedge against losing the whole data base if one drive goes down. Of course, for an extremely large data base, a 256 disk box may make sense. However, for the average user, I think the 50 to 100 disk box looks best.

For smaller data bases, say in the category of 10 gigabytes, it may make sense to use single drive units. This could then eliminate the need for a buffer magnetic disk. However, in the actual operation of having any number of users looking for data, I believe a magnetic buffer is a necessity (Exhibit 1). However, a user today will wait a day or two for a print because that is the norm. He or she can be doing something else until it gets there. But when a system such as this is installed, that same person sits at a terminal or a hard copy printer and a delay of even a couple of minutes (instead of days) will make that person unhappy. While retrieving data from an optical disk is fast, it is not as fast as from a magnetic disk. And because the optical disk system has the entire file, it has to be free to keep accessing its file, not wait while someone is looking at or working on a specific drawing. As an example, suppose we ask for drawing 123 which has 92 pages to it. Without a buffer, this could easily bring the system to its knees. Instead, all 92 pages of drawing 123 are transferred to a magnetic disk buffer, the first page is immediately transferred to its requester's CRT while the rest of the pages are being transferred to the magnetic disk to be accessed by the user as he or she sees fit. The optical disk is then free to answer other requests. This magnetic buffer should be able to handle at least 250 drawings. If it becomes overloaded, another magnetic disk buffer can be added.

4. OUTPUT ASPECTS

The super system will need several types of output capabilities also.

- o CRT – Alphanumeric only
- o CRT – Graphic and Alphanumeric
- o CRT – Graphic and Alphanumeric with change capability
- o Hard Copy Printers
  - – "A" size printer
  - – "C" size printer
  - – "E" size printer
  - – Laser printer (up to "D" size)
  - – Computer output microfilm (COM)

The alphanumeric only terminals will be placed in various managers' offices and in strategic locations within the engineering and manufacturing areas. They will be used to access status data and request that a drawing be printed out at a specific location.

Graphic terminals will be placed in areas where drawings are normally requested: drawing offices, production control offices, engineering locations, etc.

The graphic terminals with change capability will be placed in the drafting locations. Only authorized users will be able to access the change capability.

The "A", "C", and "E" size printers will be placed in strategic locations in all functions. In our areas, we will be placing about 26 terminals and 16 printers (Exhibit 2). Because it would be inefficient to print all "A" size on "C" or "E" size paper, we will have an "A" size printer available at every graphics terminal in addition to a possible "C" or "E" size printer.

A laser printer is faster and prints on regular bond paper. Therefore, a laser printer will be placed in the central drawing file location and be used for batch and multiple printing. Any terminal may direct that a copy be printed on the central laser printer, therefore an address or location notation should be included.

The COM must be an electron beam device rather than the usual CRT type COM to achieve the resolution needed for "D" and "E" size drawings.

Alternately, the CRT type COM could be used for "A", "B", and "C" size drawings and the laser flat bed printer for the "D" and "E" sizes. The "D" and "E" size prints would then be photographically reduced to 35mm film. This latter approach has the advantage of providing higher resolution at a much lower cost.

5. INTEGRATION TO CURRENT PROCEDURES

Exhibit 3 shows the drawing distribution system as it exists today. Exhibit 4 shows the procedure after the system has been installed. Tracing storage is eliminated, all satellite files are eliminated, assuring a single controlled data base. Even the microfilming of drawings is eliminated using the COM to provide aperture cards on demand.

The system also allows transmission to remote sites many miles away from the central system. The remote satellite site will have a micro system consisting of an optical disk drive or two. What took days to get a drawing to that plant can now be done overnight. Drawings directed to the remote site will be queued and when the telephone lines are free at night, they are transferred from the central location to the remote site. Exhibit 5 shows the times to transmit a document based on a 20:1 compression. As you can see, it is not too lengthy and even a 10:1 compression is viable. The use of a high speed line at 56000 baud obviously speeds things up significantly. However, the cost is prohibitive and difficult to justify. A satellite transmission to a foreign country is also possible and could save money in the long run. However here there is a problem with customs in some countries.

6. CONCERNS

While all of this looks ideal and is possible, there are some concerns which must be considered. Ask most people involved in the optical disk business and they will tell you their main concerns are the media and the read-write drive. I felt this way at first but now am convinced that this is no longer a concern to me. There are so many companies involved with the media that some are bound to work. Should the optical disk bomb, an unlikely prospect, other media such as magnetic optical technology are coming along.

But there are real concerns, the first being the software. Software is the life's blood of this system. It just has to be right. Many companies with excellent ideas and good hardware designs have gone down the drain because of software problems.

Other concerns have to do with the operational aspects. I discussed the great advantages of the change capability. Unfortunately, there is a hooker which we are still trying to resolve. It has to do with the terminal itself. The problem is easy to see (Exhibit 6). Note that a terminal with a resolution of 2200 x 1700 LPI can show an "A" size drawing at 200 LPI (the scanning rate). However, "E" size drawings require 8800 x 6800 pixels to achieve a resolution of 200 LPI. But with only 2200 x 1700 pixels available on the terminal, all we can see is 1/16 of the pixels or effectively 50 LPI.

Incidentally, following this same reasoning, a "C" size drawing will be viewed at 100 LPI at the terminal. I am not sure if a drafter can see enough at 100 LPI to make a change, certainly not at 50 LPI. To achieve 100 LPI on an "E" size drawing, you will be able to review 1/8 of the drawing (1/16 at 200 LPI). Incidentally, here is what an "R" looks like at 200, 100, and 50 LPI (Exhibit 7). It will be difficult to make changes under these circumstances and, as I said, we are still trying to resolve this problem by providing dynamic zooming and panning controls for viewing portions of the drawing at higher resolutions with rapid access to adjacent areas. As another consideration, if you are looking at an "A" size drawing in vertical (or portrait) format and the screen is in a horizontal (or landscape) format, then the 2200 LPI must fit into the 1700 pixel availability, reducing the document to 77 percent of size with a resolution of about 154 LPI. By the way, this only applies to looking at the drawing on a terminal; printing it out on an electrostatic or laser printer will result in resolution as scanned—200 LPI.

Another concern is the actual operation. Asking some people in manufacturing or engineering to use a CRT and read a print from the screen instead of hard copy will cause fear and trepidation among the ranks. Telling a drafting manager that from now on all his drawings will be on some unreadable (to him) disk will send him into an emotional state of anxiety. Because this is new technology, it will be met with distrust and, unless the system is carefully planned, they (the users) will make it fail. They have to be brought into the picture early in the planning stages to get their support. One cannot go to the Manager of Manufacturing and tell him that the great system will eliminate 20 of his people—he will throw you out. Everyone tends to be protective of their own. But with a proper approach, everyone will be behind it and it will go.

Which brings up the next concern. Once all of these people are enthusiastic, the system will be utilized to its full extent. However, if the system is not designed adequately to meet the demand, then delays will occur and their enthusiasm will turn to frustration.

The person in charge of the system will have to be a very dedicated worker and skillful negotiator. A successful system will tend to grow and thus the system must have the ability to be upgraded—a modular approach is a prime requirement. I can see 20 or 30 people, all accessing information at relatively the same time. An occasional wait because of a queue will be tolerated but not a constant delay. That optical disk drive will be a busy little device.

And what happens when it's down? It will fail at the worst possible moment. A spare drive, if not a juke box, is a necessity. This system will be placed in a dynamic environment and it must be equal to the task.

When it works, it will have an excellent payback—even more than we can identify today. The system can be justified as Medium Steam has done (Exhibit 8). These savings are conservative and I think will exceed twice what has been identified.

7. CONCLUSION

In spite of all these concerns, are there systems today that will provide what we want or is this just fantasy?

For the most part, this system can be available today. There are several major suppliers that were looked at which are possible sources and more suppliers are just announcing or are about to announce systems.

In GE, the four systems looked at were:

ALPHAREL
INFODETICS
INTEGRATED AUTOMATION (TCI)
TIECO (TERA)

All of these profess to be in the optical storage disk business. However, Infodetics stresses the aperture card storage system and TIECO stresses roll film, either 16- or 35-mm.

In our study, we did not fully appreciate the complexity of designing a system such as I have described. In the past several weeks in consulting with ALPHAREL, I gained an understanding of the design sophistication and experience required to bring a viable system to the market place. Problems which at first glance appear insignificant can be a major stumbling block. For example, during the scanning of an aperture card, the orientation of the aperture is critical. If it's skewed or poorly mounted, it is possible some data could be missed. Therefore, the system must be designed to accommodate these inconsistencies. This is only one example. There are many more with more severe complexities.

By 1984, some excellent systems will be on the market in spite of the complexities. I just want to impress you with the required sophistication in this new technology.

It's a system we badly need if we are to survive in the area of drawing reproduction and distribution.

8. BIOGRAPHY

Mr. Gingrich, after graduating from Lafayette College in Easton, Pennsylvania, joined the General Electric Company where he spent most of his 35 years in various engineering department services work. In 1968, he accepted the position of Manager, Engineering Services and Administration in the Medium Steam Turbine Department. Among other duties, he was responsible for all of the functions associated with drawing and technical manual reproduction. He initiated, procured, and managed the first interactive graphics system for the Medium Steam Turbine Department. He later participated on the GE Corporate study team on Data Base Management Systems and led the study team in automated drawing storage and retrieval systems. As a result of these studies, one General Electric Department has placed an order for an optical disk storage system with many others to follow. In June 1983, Mr. Gingrich retired from the General Electric Company to go into consulting work and has recently agreed to act as a consultant to ALPHAREL, Incorporated.

EXHIBIT 1

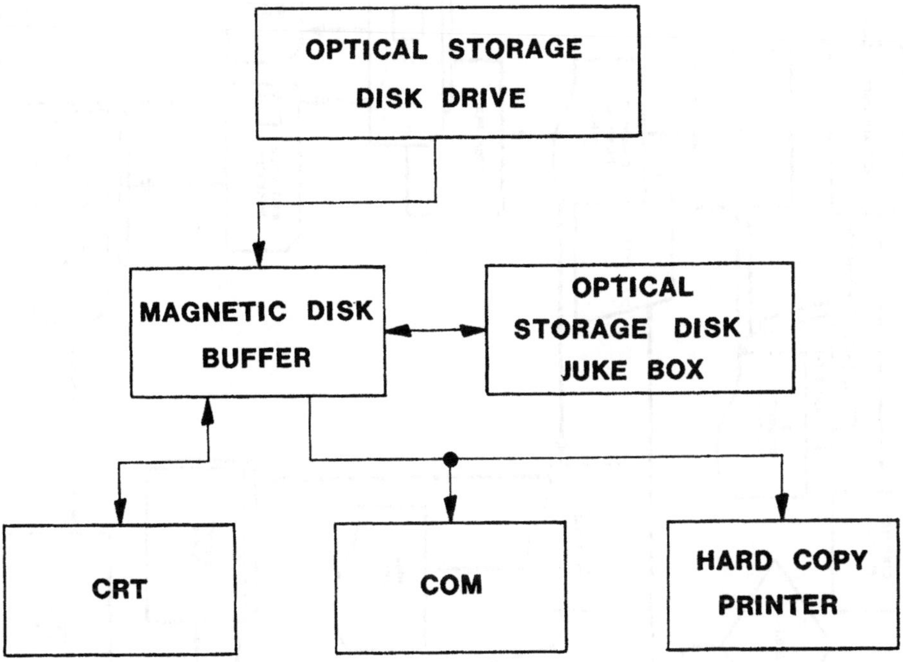

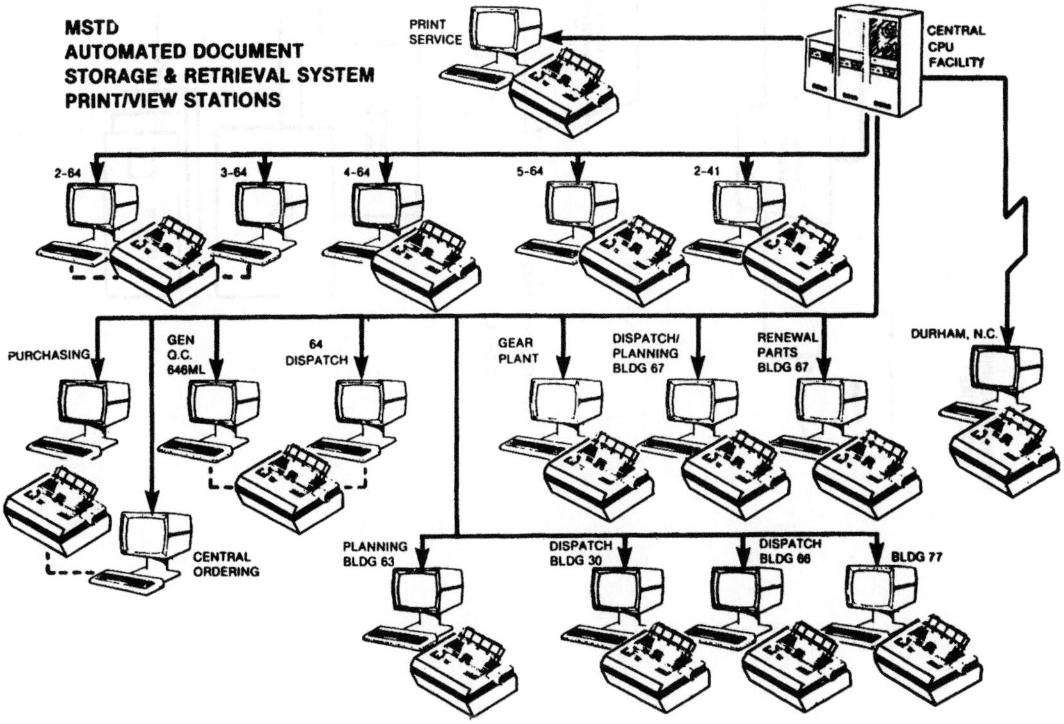

EXHIBIT 2

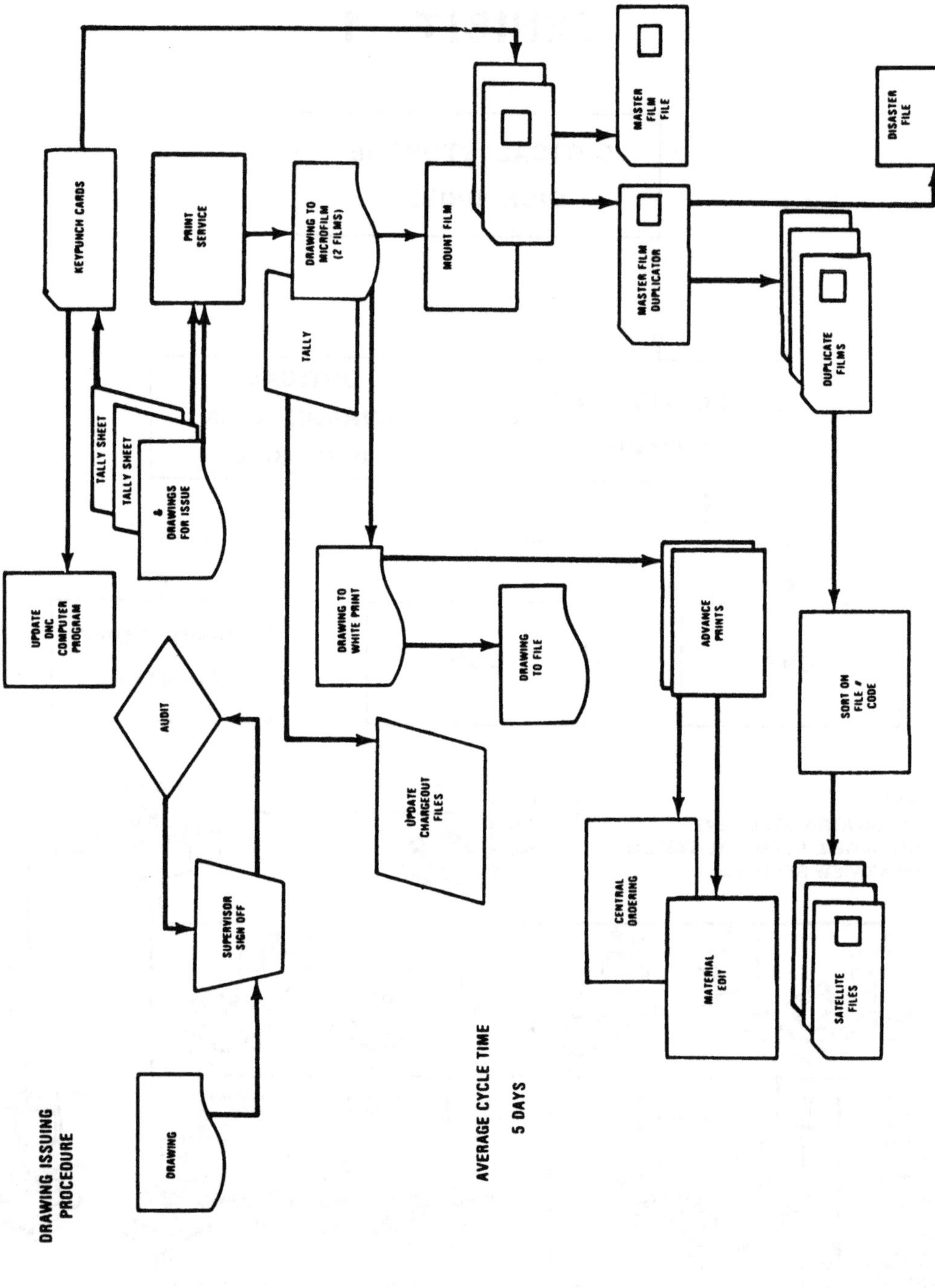
DRAWING ISSUING PROCEDURE
DRAWING
SUPERVISOR SIGN OFF
AUDIT
UPDATE DNC COMPUTER PROGRAM
TALLY SHEET
TALLY SHEET
& DRAWINGS FOR ISSUE
KEYPUNCH CARDS
PRINT SERVICE
DRAWING TO MICROFILM (2 FILMS)
TALLY
MOUNT FILM
MASTER FILM FILE
MASTER FILM DUPLICATOR
DISASTER FILE DURHAM, N C
DUPLICATE FILMS
DRAWING TO WHITE PRINT
DRAWING TO FILE
UPDATE CHARGEOUT FILES
ADVANCE PRINTS
CENTRAL ORDERING
MATERIAL EDIT
SORT ON FILE # CODE
SATELLITE FILES
AVERAGE CYCLE TIME
5 DAYS

EXHIBIT 3

EXHIBIT 4

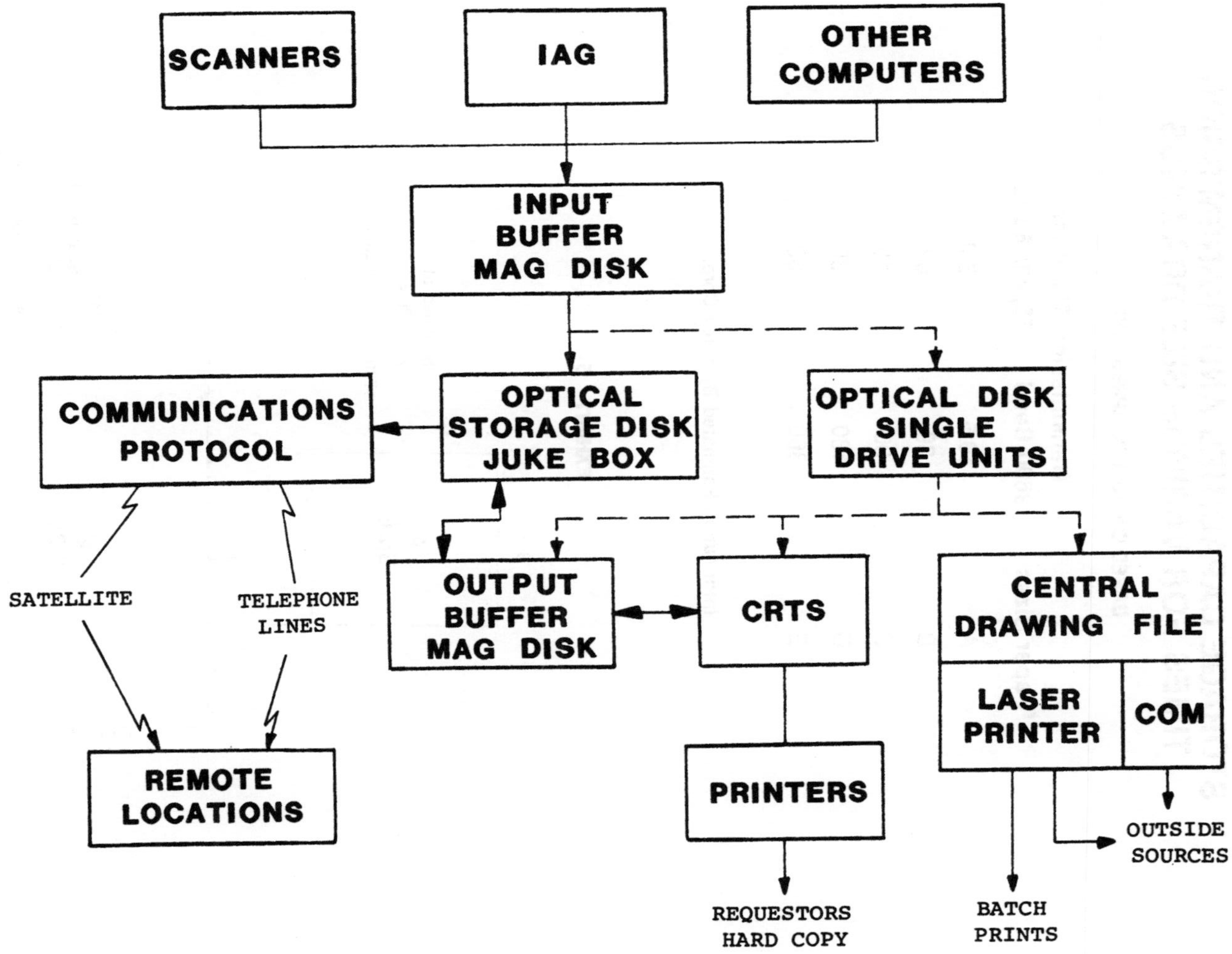

SCANNERS
IAG
OTHER COMPUTERS
INPUT BUFFER MAG DISK
COMMUNICATIONS PROTOCOL
OPTICAL STORAGE DISK JUKE BOX
OPTICAL DISK SINGLE DRIVE UNITS
SATELLITE
TELEPHONE LINES
OUTPUT BUFFER MAG DISK
CRTS
CENTRAL DRAWING FILE
LASER PRINTER
COM
REMOTE LOCATIONS
PRINTERS
OUTSIDE SOURCES
REQUESTORS HARD COPY
BATCH PRINTS

STORAGE CAPACITIES AND TRANSMISSION TIMES FOR VARIOUS SIZE DRAWINGS

BASED ON 20:1 COMPRESSION

	Transmission Time/Sec.	
Paper Sizes	9600 Baud	56,000 Baud
A	19.5	3.2
B	39	6.5
C	90	15
D	180	30
E	360	60

Information Estimated By Xerox Corp.

EXHIBIT 5

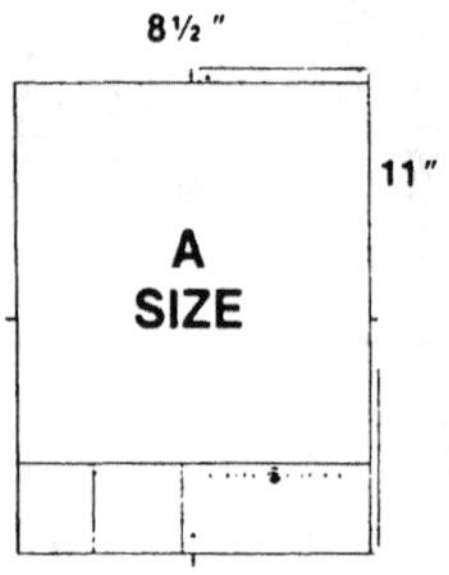

Scanned at
200 LPI =
1700 × 2200 Pixels

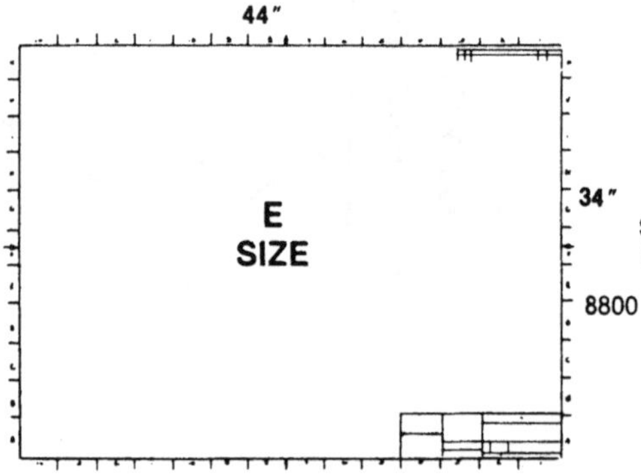

Scanned at
200 LPI =
8800 × 6800 Pixels

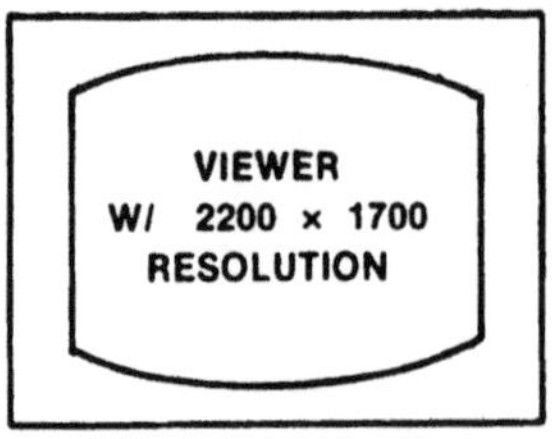

A Size will be seen at 200 LPI Resolution

E Size 2200/8800 = 1/4 of 200 LPI or 50 LPI

EXHIBIT 6

READABILITY

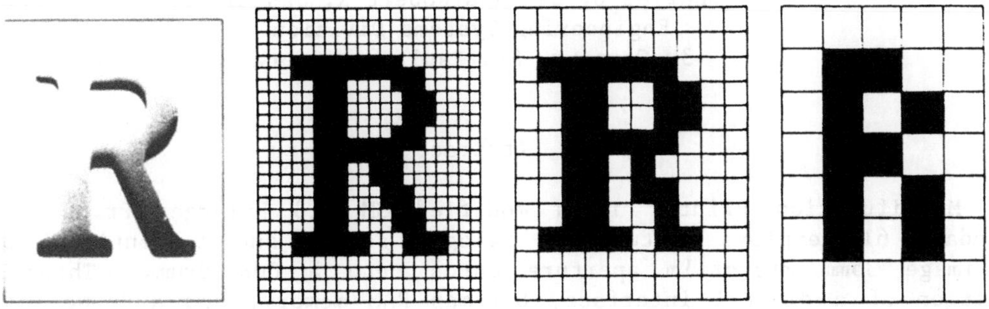

200 LPI　　100 LPI　　50 LPI

EXHIBIT 7

OTHER SAVINGS

Eliminate Outside Vendor Services
Eliminate Shop Loss from Unreadable Drawings
Eliminate DNC File from Honeywell
Cancel DNC On Line Development
Eliminate Miscellaneous Other Comp. Reports
Eliminate Machine Maintenance

PEOPLE - SALARIES AND BENEFITS

4 Grade 3 Clerks - Print Service
2 Grade 3 Clerks - Drafting Services
1 Level 5 Supervisor - Print Service
6 Drafters (Avg DO-1)
3.5 Production People
Overtime Reduction

PAPER AND SUPPLIES

40% Reduction in Prints
Eliminate CPL Film and Print
Eliminate Satellite Files
Eliminate Pre-Print Tallys
Reduce Print Mach Supplies
Photolitho Reduction

MACHINE RENTALS

2 Quantimatics
1 Xerox 1824
2 OCE 3600
1 Xerox 600

EXHIBIT 8

14

3M MULTIFUNCTION PRINTER FOR CAD

Charles C. Lee and Robert R. Jasper
Engineering Systems Division
3M Company, St. Paul, Minnesota

Abstract

The Multifunction Printer is a modified 3M 618 Enlarger Printer. The standard 618 employs a tungsten halogen source and conventional optics to image 35mm microfilm aperture cards onto a CdS drum. Through the addition of a digital interface, bit-map RAM memory, and a He-Ne scanner, the modified 618 can print from microfilm and from the digital vector files of a CAD host system. Currently the mechanics of the imaging system allows for sizes of graphics output up to 18 x 24 inches which encompasses the common C-size (17 x 22) engineering drawing format. The 618 printer engine employs the KIP electrophotographic process and utilizes a two-part, heat fusing toner to form images on plain bond paper, vellum, and other media.

1. INTRODUCTION

The 3M Multifunction Printer consists of three main subsystems -- the laser scanner and the printer engine, and an electronic interface with I/O Board. It provides wide format prints up to 18 x 24 inches at a paper speed of 5 inches/sec. It can print either from computer digital data or from microfilm. The mode of operation is operator selectable by switches. Applications for this product would be 1) CAD printing/plotting, 2) An output device for micrographics scanning and storage systems or optical disk based systems, 3) Aperture card file locations that require CAD output capability.

2. 3M MULTIFUNCTION PRINTER PROCESS

The 3M Multifunction Printer Engine, as shown in Figure 1, is an electrophotographic printer that uses a laser to plot graphic electronic data from a CAD system, and it uses a conventional projection light source to make prints from a microfilm aperture card. It accepts sheets of plain paper 16 to 20 pound bond, vellum and offset masters. The maximum print size is 18 by 24 inches, which encompasses the common C-size (17 x 22) engineering drawing paper. The printer engine is 3M 618 which employs the KIP electrophotographic process and utilizes a two-part dry toner to form images on a CdS drum. As illustrated in Figure 2, the printing process utilizes a photoconductive CdS plate wrapped around a drum. On top of the CdS coating, there is a thin insulating layer.

In operation, the photoconductor plate rotates past a number of treatment stations to form a transferable image. The sequence begins as the plate passes under the first corona, where it receives a positive charge to erase any previous image and establishes a uniform positive potential on the surface. The plate next passes the negative corona, where it is simultaneously imaged from the projected microfilm or from the modulated laser beam. The light struck area of the photoconductor becomes conductive and allows the polyester surface to accept the negative charge, while the dark areas repel the charge and remain positive. Following the negative corona is a post-expose lamp, which has the effect of balancing the charges across the polyester film and fixes the electrostatic image on the plate surface. The plate now passes the developing station, where the developer, consisting of a positively charged toner powder and an iron carrier, is brought into contact with the plate surface. A magnetic brush removes excess developer, leaving an image of toner attracted to the negative electrostatic image. The imaged plate then comes in contact with plain paper, and the toner is transferred to the paper with the help of the transfer corona and fused by heat. The previously imaged plate next passes a cleaning station to remove residual developer before beginning the cycle again. Because the black images are developed from the light striking area, high quality and extremely low background prints can be obtained.

3. LASER SCANNER SUBSYSTEM

Figure 3. illustrates the laser scanner subsystem. A 7 mW helium-neon laser is the light source for electronic printing. A beam compressor reduces the 1mm diameter laser beam to about 140 micron beam waist located at the center of the acousto-optical modulator (AOM) to minimize the rise time. The beam expander that follows the AOM is similar to a Galilean telescope which enlarges the Gaussian beam to the required size to produce a diffraction limited spot size focused on the surface of the photoconductor drum. An 18 facet polygonal mirror is used to scan the beam horizontally parallel to photoconductor drum axis. A pair of cylindrical and toroidal anamorphic lenses was used to reduce the beam position errors at the drum due to facet to facet tilt angle error. A flat field f-θ lens is required to produce the scanning length of 18" and resolution of 250 dots/inch. Folding optics are needed for containment in the optics box. A start of scan detector is used to provide a correctly timed signal to start writing for each line thus to reduce the polygon facet error to insignificance. The facet width of the polygonal mirror is chosen to be twice the diffraction limited aperture diameter so that a 50% duty cycle operation can be achieved. This is the trade-off between conserving the laser energy and minimizing the dimension of the polygonal scanner mirror The scanner motor has phase locked loop control while maintaining its speed to 0.1% per minute, and the high inertia prevents rapid motor speed fluctuation. In Table I. the specifications for the laser scanner subsystem are summarized:

TABLE I.

Laser HeNe	7mW @ 6328 nm
Gaussian Spot Size At Image Plane	4 mil
Flat Field Scan Length	18 inches
Duty Factor	50-60%
AO Modulator	Isomet IMD-80
Scan Rate	1250/sec
Data Rate	11.25 MHz

4. ELECTRONIC INTERFACE

When the Multifunction Printer is operated in computer mode, the electronic data has to be rasterized and presented to the printer as a bit map at a data rate of 11.25 MHz, which is required by the printing process. However, if the data is in vector form and cannot match the speed of the printer, then a vector-to-

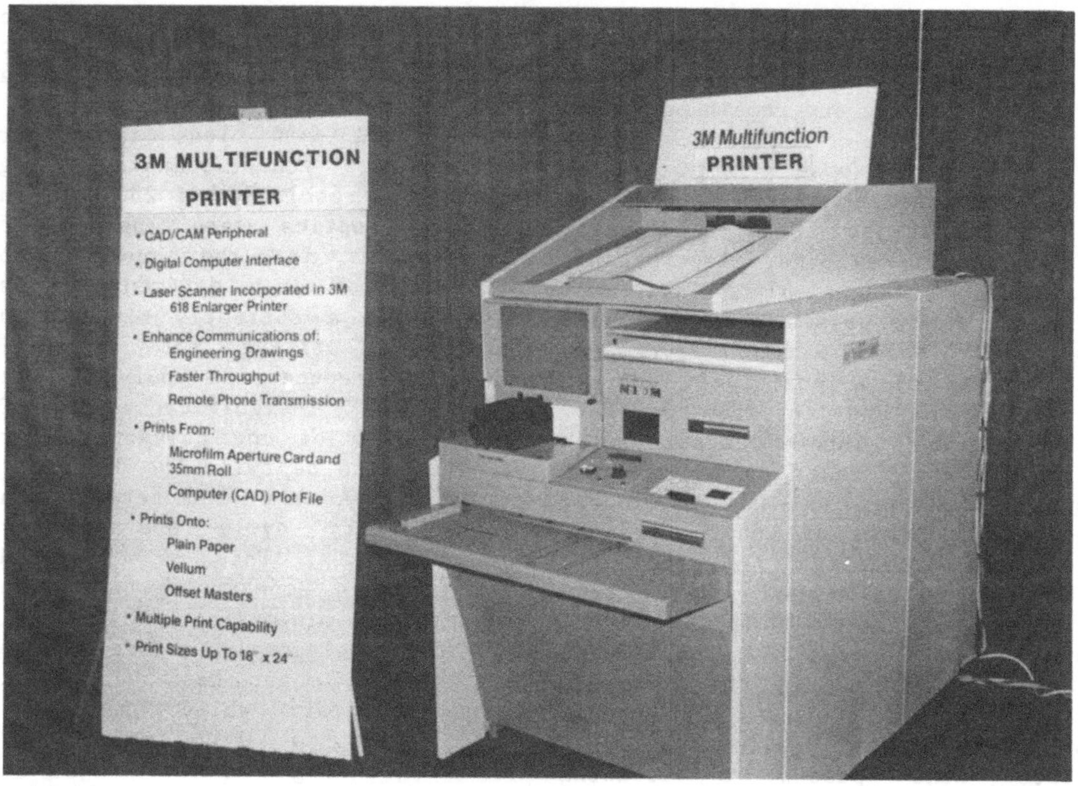

Figure 1. 3M Brand Multifunction Printer

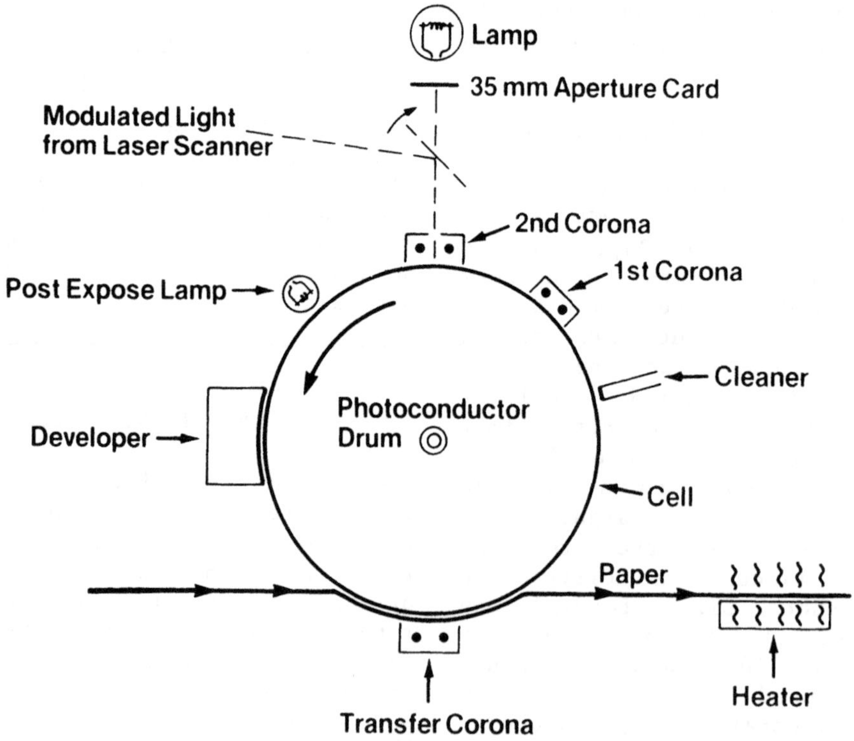

Figure 2. Printing process

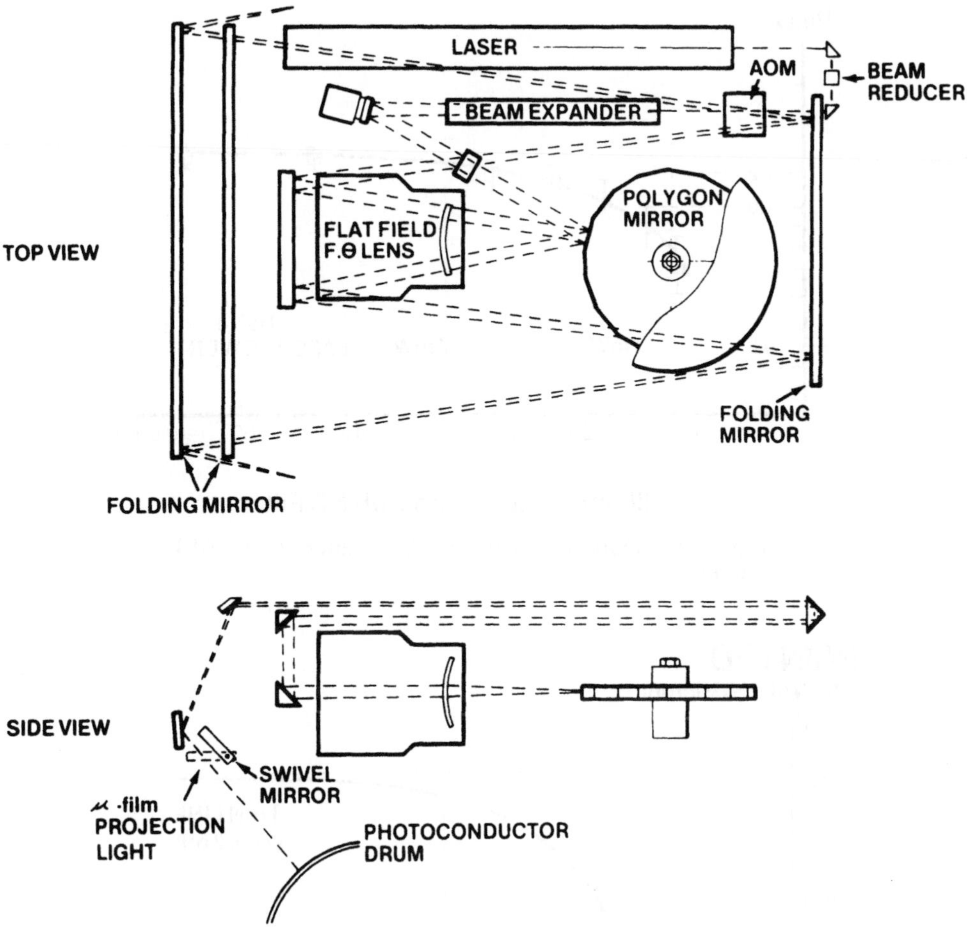

Figure 3. Laser scanner subsystem

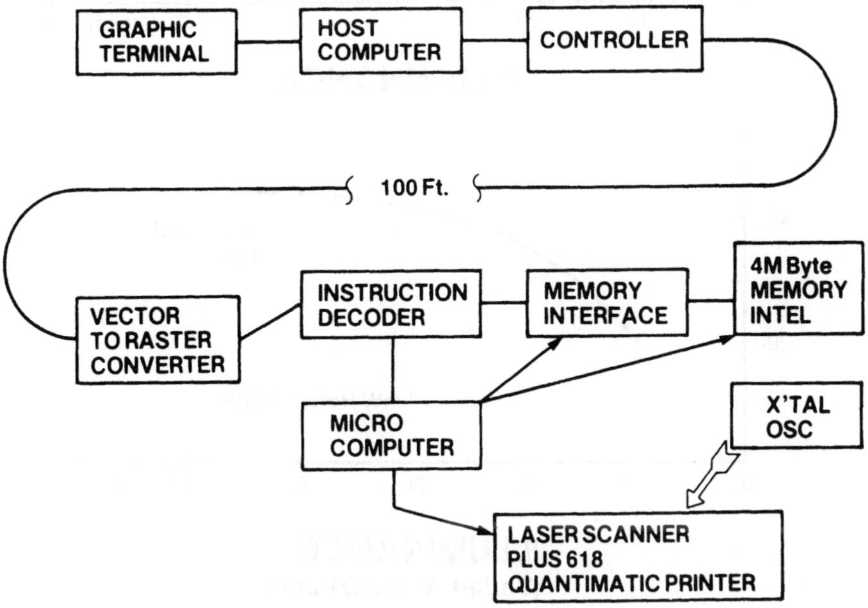

Figure 4. Vector input electronic interface

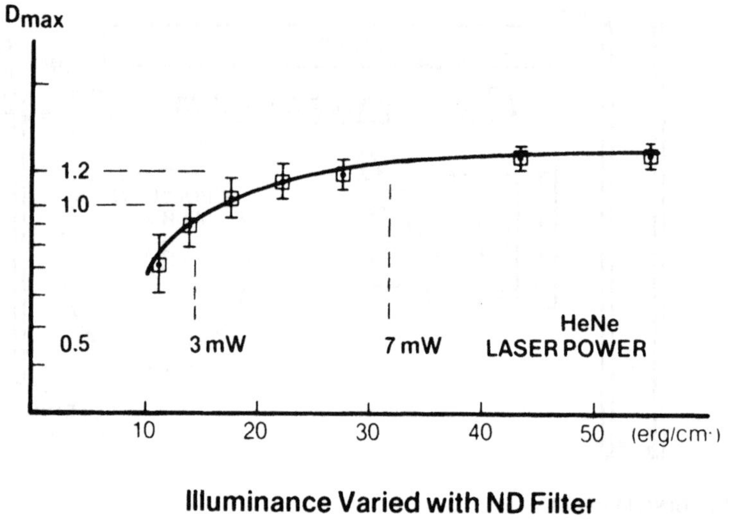

Figure 5. Optical density vs. light intensity in 618 printer

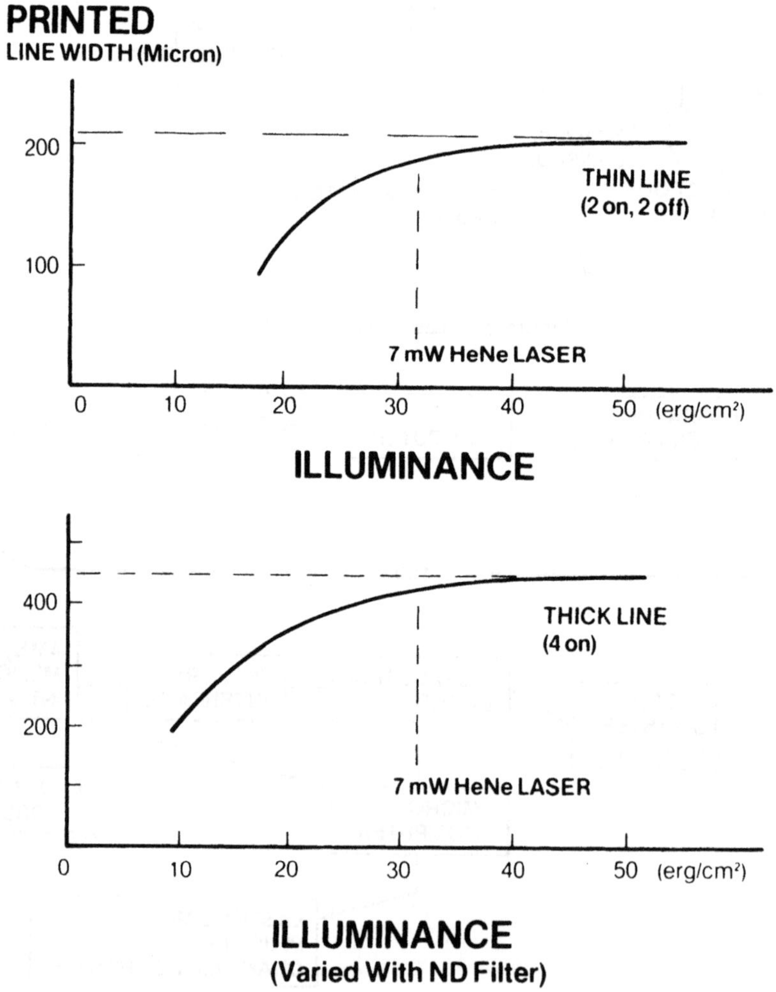

Figure 6. Printed line width vs. illuminance

raster processing either by host computer CPU or by a stand-alone unit is needed, as well as a 4 Megabyte page memory buffer. Fig. 4 illustrates this mode of operation. A parallel I/O was hard wired between the host computer controller and the vector to raster converter in order to achieve high speed data transmission. The rasterized data is then used to load the 4 megabyte page memory which consists of 64K Dynamic RAM. Once the buffer is loaded, the graphic data in serial form was clocked out by a crystal oscillator synchronized by the start of scan pulses, and fed to AOM to provide laser modulation.

5. PROCESS CHARACTERIZATION

The printing quality of the output from the Multifunction Printer operated in Laser mode was evaluated using a reflection and scanning densitometer. The laser intensities in terms of ergs/cm^2 were varied by means of different neutral density filters, placed across the light beam just before the photoconductive drum. As shown in Fig. 5, the maximum optical density in the dark area is plotted against the original laser illuminance. A 7mW laser power is required to generate black image having optical density of 1.2. Similarly, the printed line width versus laser illuminance is shown in Fig. 6 for the case of thin line (two black lines and two white lines) and for the case of thick line (four black lines and four white lines), respectively. It should be pointed out that the line width indeed increases with the laser power, but it apparently tends to level off at 7mW, which corresponds to 3 to 4 mW power level on the surface of the photoconductor. The HeNe laser being rated as Class IIIB is contained within the scanner box in the printer, which has safety interlocks to prevent any exposure leakage to the user. The Multifunction Printer is thus a Class I product.

5. BIOGRAPHIES

Charles C. Lee

Dr. C. C. Lee received the M.S. and Ph.D. degrees in solid state physics from Purdue University in 1974 for laser non-linear optics. He joined 3M in 1974 where he has worked in the Systems Research, Physics and Materials Laboratories, Duplicating Products Laboratory, and Engineering Systems Laboratory. His experience is in the area of electrophotography; materials and process research; laser scanning, and electronic marking technologies.

Robert R. Jasper

Mr. Jasper is a project leader in the Engineering Systems Division of 3M Company. He earned a B.S. degree in Photographic Engineering from Rochester Institute of Technology in 1971. After spending three years as a project manager in the U.S. Air Force Tactical Air Warfare Center, Mr. Jasper joined 3M as a process engineer in the Photographic Division. In his current position with the Engineering Systems Division, Mr. Jasper is responsible for development of wide format plain paper printers for the Engineering marketplace.

15

THE NATIONAL COAL LIQUEFACTION TECHNOLOGY DATA BASE SYSTEM

George W. Pukanic and William C. Peters
Department of Energy
Pittsburgh Energy Technology Center
P. O. Box 10940
Pittsburgh, Pennsylvania 15236 U.S.A.

Abstract

Procedures have been established for inputting, processing, and retrieving large volumes of technical data and backup engineering information archived in the Department of Energy's National Coal Liquefaction Technology Data Base (LTDB) System, located in Pittsburgh, PA. A Technical Inventory System has been established for accounting and classifying all useful technical information relating to coal liquefaction technology. This represents around 1.5 billion dollars for research, development, and demonstration efforts, largely government sponsored, during the major funding period from 1977 through 1983. An Engineering Summary/Data Compilation Management System has been designed and established for processing and summarizing the engineering data classified as high quality/high utility from an engineering design perspective. Several search routines developed as part of the management information system approach for the LTDB are presented.

1. INTRODUCTION

Around 1.5 billion dollars have been spent on funding coal liquefaction technology research projects during the seven year period from 1977 through 1983 as a result of the fuel crisis prevalent in the last decade. Figure 1 gives the funding profile for the four major liquefaction projects: SRC-II, SRC-I, H-Coal, and EDS.* With the current alleviation of the oil crisis, there has been a correspondingly decreased emphasis on synthetic fuels programs, and a considerable reduction of government and industry support for synthetic fuels-related projects. It has therefore become imperative to preserve and manage the vast amount of information accumulated as a result of these research efforts for future research needs that could result from a revival in a national synfuels program.

As the lead national Energy Technology Center for coal liquefaction technology, the Pittsburgh Energy Technology Center (PETC) has the responsibility of establishing a liquefaction technology data base (LTDB) for all

---

* *SRC-II* refers to the solvent-refined coal liquefaction process initiated by the Pittsburg and Midway Coal Mining Company. This process liquifies coal with hydrogen donor solvent under hydrogen pressure and at elevated temperatures, and includes bottoms recycling. The primary product is a liquid distillate fuel. *SRC-I* is similar to the SRC-II process, but due to less severe operating conditions, the principal product is a solid solvent-refined coal that is almost ash free. *H-Coal* is a process developed by Hydrocarbon Research Inc., for liquifying coal through hydrogenation of the coal in an ebullated-bed catalytic reactor. *EDS* refers to the Exxon Donor Solvent Process developed by Exxon Research and Engineering Company, which liquifies coal with a recycle solvent that is catalytically hydrogenated external to the reactor.

Fossil Energy liquefaction projects. The objectives of this effort are to develop a data base management system (DBMS) to collect and organize the existing and future coal liquefaction-related data from industrial, government, and foreign sources; to provide state-of-the-art information and shorten the lead time for future development of synfuels industries; to provide lessons-learned cross-cut analyses of advanced coal liquefaction processes; and to provide a baseline for comparison of advanced coal liquefaction technology research.

Arrangements have been made to deliver the following information from all Fossil Energy coal liquefaction projects to PETC: physical, chemical and thermodynamic (PCT) properties of coals and coal liquids; plant and equipment design, costs, construction, operation, and performance; properties of materials of construction; engineering development and analysis studies (e.g., data correlation, computer simulation, and parametric studies); commercialization technology assessment; health, safety, and environmental aspects; government policy and regulations; and administrative and management aspects. Information from the SRC-II coal liquefaction project has been received and consists of approximately 800 cubic feet of hard copy, 5 cubic feet of microfiche, and 1000 drawings on 35-mm aperture cards. Similar volumes of data are expected to arrive in the near future from the SRC-I, H-Coal, and EDS coal liquefaction projects.

At present, there are two information management systems in use at PETC for the liquefaction technology data base system. The first, called the Technical Inventory System, contains an inventory of the technical content of all documents relating to liquefaction technology* and is written in the Fortran programming language. The second, called the Engineering Summary/Data Compilation Management System, contains engineering summaries of the most useful of these documents, referred to as the high-quality/high-utility information from an engineering design perspective; this system utilizes the VAX-11 DBMS, which is a CODASYL-compliant, general purpose data base management system.

2. INFORMATION PROCESSING PROCEDURE

Information received is processed through essentially five work stations, as depicted in Figure 2. The incoming information is first sent to Station I, where non-useful information such as manufacturer's catalogs, personnel records, maintenance work schedules, etc. is removed; this information is then disposed of according to prescribed formats. The remaining useful technical information is then sent to Station II where a storage location and unique accession number are assigned to the source materials. Then, each of these documents is manually logged into the Technical Inventory System at Station III. (The information that is logged in is presented in the example search routine discussed in Section 4, THE TECHNICAL INVENTORY SYSTEM, and is displayed in Figure 6.) At Station IV, material that is deemed to be most useful to the user community, the high-quality/high-utility information, is identified and engineering summaries of this high-quality material are prepared. These engineering summaries are then entered into the Engineering Summary/Data Compilation Management System at Station V. Finally, all useful documents are checked for duplication at Station VI.

3. THE IDEAL LTDB SYSTEM

Figure 3 gives a representation of the ideal LTDB System—a goal that PETC plans to eventually implement; those areas implemented to date are so indicated. Incoming technical information exists in such forms as hard-copy documents, microfilm, engineering drawings, aperture cards, magnetic tapes, and floppy discs. This information is initially entered into the Technical Inventory System. At present, there is limited dissemination service to users; since a system is not yet fully developed for off-site user access to the LTDB, all information requests are handled either by the user on-site or by PETC

*Documents consist of technical reports, correspondence, minutes of technical meetings, laboratory and operating data, engineering drawings, process flow diagrams, computer printouts, and microfilm.

personnel; in the latter case, hard copy documentation is then sent to the user. Engineering summaries prepared for the high-quality/high-utility information are entered into the Engineering Summary/Data Compilation Management System; eventually numeric information will also be entered here. The same information dissemination limitations exist as for the Technical Inventory System.

There are some liquefaction studies being planned and others currently being carried out using information from the LTDB, as indicated in Figure 3. These efforts are discussed in Section 6, LTDB USAGE.

Figure 4 represents a design schema of input and output options, either planned or in place, for the LTDB. Technical inventory data on magnetic tapes are input directly into computer memory. Future plans include digitization of engineering drawings for electronic entry into memory. Data-entry and data-retrieval programs are currently under development. Output is in the form of hard copy printouts or a display on a CRT terminal; limited graphic display of electronic data is also available.

4. THE TECHNICAL INVENTORY SYSTEM

There are currently over 11,000 documents in the Technical Inventory System. About 80% of the information is related to the SRC-II liquefaction process, and the remainder is related to other liquefaction technology processes, such as SRC-I, H-Coal, EDS, CO_2 Acceptor, and Zinc Chloride, as indicated in Figure 5. At present, somewhat over 80% of the documents are in the form of hard copy; somewhat less than 10% are in the form of microfilm, microfiche, magnetic tapes, and photographs; and the remaining documents are in the form of engineering drawings and computer printouts.

Two search routines are presently available for information retrieval. The first is called a String Search and will search on a given string of characters (or group of strings, with each character string separated by a space), with no provision for excluding any prefixes or suffixes that may be attached to the string of characters. The second routine is called a Quick Search and will exlucde any prefixes or suffixes on a given string of characters. Both search routines can search on the intersection of character strings. In addition, the String Search can search in an "either, or" mode, i.e., on the occurrence of either one character string or another (e.g., to retrieve information on the kinetics of a reactor unit, which is sometimes referred to as a dissolver unit, the following entry using the comma would be made: reactor,dissolver kinetics). The String Search provides for the option of character string elimination (e.g., to exclude wastewater analysis from water analysis, the following entry using the asterisk would be made: water*waste analysis).

An example of a String Search, as well as the type of information available for a document, is given in Figure 6. This search represents a request made to PETC on the availability of a report for which the title and report number were known; the search was made on the report number DOE/OR/03054-T11. Since slashes and dashes are not acceptable search characters, the search was conducted using the group of strings indicated on the keyword line in Figure 6 (each character string must be separated by a space; the plus sign appears only on the output). On the next line, the following descriptive information is included: the unique document accession number, 7116; the number of copies available, 1; the storage location, box number 562; document source, ORNL for Oak Ridge National Laboratory; the process, generic; plant scale, 4 for demonstration plant; document medium, HC for hard copy; type of report, progress report; availability of abstract, N for No. The remainder of the information includes the following when available: title of the document, contract and report number, date, and table of contents.

Experience with the Technical Inventory System has revealed areas that will be the target for future improvement. Since searches have often resulted in a large number of successfully retireved documents, future search routines will provide the option of searching on a previous search to further narrow the selection in an efficient manner. Also to be implemented in the future are phrase searches, i.e., searches to insure that selected character strings fall into close proximity to each other; this would help eliminate those documents where the character strings might relate to different topics or ideas (e.g., a document that might consist of more than one report, and the character strings searched on might each be in different reports).

5. THE ENGINEERING SUMMARY/ DATA COMPILATION MANAGEMENT SYSTEM

The Engineering Summary/Data Compilation Management System presently contains approximately 200 engineering summaries of some of the most useful of the technical information, i.e., the high-quality/high-utility information, in the Technical Inventory System. These summaries generally consist of two sections: the first is more general in scope, while the second presents more details. The technical areas covered include the following: physical, chemical, and thermodynamic properties of coal liquids; plant and equipment design, construction, operation, and performance; engineering development and analysis studies (e.g., data correlation, simulation, and parametric studies); technology assessment for commercialization; and environmentally related subjects.

Figures 7 and 8 present the basis for this management system. Figure 7 gives the LTDB classification structure breakdown to three levels: liquefaction process type, plant area groupings, and plant areas. Figure 8 gives the LTDB indexed classification structures, which includes the levels in Figure 7 as well as a further structure breakdown to include plant scale, the specific unit operations (only seven are shown), the data category, and the data status. Each engineering summary is classified according to the appropriate category/ies in Figures 7 and 8.

In the Technical Inventory System, a search is carried out on character strings that make up the text of the information stored for a given document (i.e., a search is carried out on titles, tables of contents, or any other descriptive technical information that has been entered for a document). In the Engineering Summary/Data Compilation Management System, however, a keyword search is carried out on keywords or keyword combinations that are entered separately with each engineering summary; no search is made on character strings that make up the text of the engineering summary.

An examble of a search session is given in Figure 9. At the beginning of a session, the user has a choice of searching on a keyword/s, accession number, title, index number, author, or company. The example search in Figure 9 represents a search on the keyword ENVIRONMENTAL. Since this keyword is not in the data base, the system searches the Thesaurus.* Three phrases containing the entered keyword have been found, and the user can now choose to search the data base on any of these keyword phrases. In the example, the choice has been made to search on ENVIRONMENTAL REGULATIONS. Since only one document has been found, the engineering summary (as well as title, author, etc.) is displayed; if more than one document were found, the titles of up to ten documents would be displayed, since ten documents constitute a so-called "working set" (additional documents would constitute additional working sets, and an option exists for displaying these). The user would then have the option of displaying a particular engineering summary (Figure 10 gives an example of a multiple title listing; note truncated titles due to line size limitations). Due to CRT computer terminal screen limitations, no more than nine lines of engineering summary text are displayed at one time. As shown in Figure 9, the option exists for displaying any additional pages of text.

Near-term plans involve expanding the data base to include integration of the numeric information (particularly relating to the very important physical, chemical, and thermodynamic properties of coals and coal liquids) with the technical information. Longer-range plans involve development of graphical display capabilities in an interactive mode.

---

*The Thesaurus at present does not contain synonyms; if the entered keyword phrase is not in the data base, the Thesaurus is searched for those phrases most closely matching the entered phrase.

6. LTDB USAGE

Although considerable information exists on PCT properties of petroleum liquids, there is generally insufficient information on PCT properties of coal liquids. Yet such information is required to carry out process simulation studies on coal liquefaction systems. A considerable amount of data is available in the PETC LTDB that can be used to provide PCT properties information on coal liquids. As briefly mentioned in Section 3, which refers to the ideal LTDB system as depicted in Figure 3, there are currently some liquefaction analysis studies in progress that use data available in the LTDB. One such study uses some basic physical property characterization parameters from the LTDB to develop the required PCT property correlations. Another study makes use of these correlations as well as other process data from the LTDB to simulate a coal liquefaction process using the ASPEN system, which is a process simulator designed for analysis of liquefaction systems. Figure 11 indicates how these modelling efforts utilize information in the LTDB and also serve to validate existing data; improve existing, and develop new, models; and add additional information to the LTDB.

As indicated in Figure 3, cross-cut studies are planned among the major liquefaction processes, and such studies would make use of information in the LTDB. Cross-cut analysis could be made of process design and economics, critical equipment performance, mechanical problems, process operational and maintenance aspects, and feedstock and product slates. Any future synthetic fuels plant would benefit from such analyses and from the lessons learned in the developmental efforts for the major liquefaction processes.

After being fully developed, the LTDB will provide the user with a search capability of updated state-of-the-art information on coal liquefaction systems. In addition to providing bibliographic and textual information, the LTDB will have a considerable numeric data base. When completed, the PETC LTDB will contain the enormous amount of technical information generated from the comprehensive research and development programs in coal liquefaction. It will be in a form that can serve as a basis for comparison of advanced liquefaction technology research, as well as for a new synthetic fuels industry in the future.

7. DISCLAIMER

Reference in this report to any specific commercial product, process, or service is to facilitate understanding and does not necessarily imply its endorsement or favoring by the United States Department of Energy.

8. BIOGRAPHIES

Dr. Pukanic is a technical information specialist/physical scientist in the Office of Technical Information Management at the Pittsburgh Energy Technology Center. He is responsible for the development of the management information systems for the National Coal Liquefaction Technology Data Base. In addition to the planning and design of these systems, he is responsible for incorporation of the technical information into the data base system.

Dr. Pukanic received is Ph.D. in Chemistry from Duquesne University and his M.S. degree in Industrial Engineering: Operations Research and Systems Management from the University of Pittsburgh. He has extensive experience in modelling and simulation activities in coal gasification and liquefaction systems. In his extensive project management activities, he has been technical project officer for a number of projects in the coal conversion technology area. He was responsible for the development of a data base management system for contracts monitoring in the Advanced Research and Technology Development Program for Direct Utilization of Coal and Synthetic Fuels in the Office of Fossil Energy.

Dr. Peters is a supervisory physical scientist responsible for administration and technical management of the Office of Technical Information Staff at the Pittsburgh Energy Technology Center, DOE. Responsibilities since 1980 have principally involved the development of the National Coal Liquefaction Technology Data Base (LTDB) from inception to fruition. Currently, Dr. Peters is managing the adaptation of the LTDB system to other coal technology areas such as Conventional Coal Combustion, Coal-fired Magnetohydrodynamics (MHD), Coal Water Mixtures, and Flue Gas Cleanup.

Figure 1.
Coal Liquefaction Appropriations for Major Liquefaction Projects

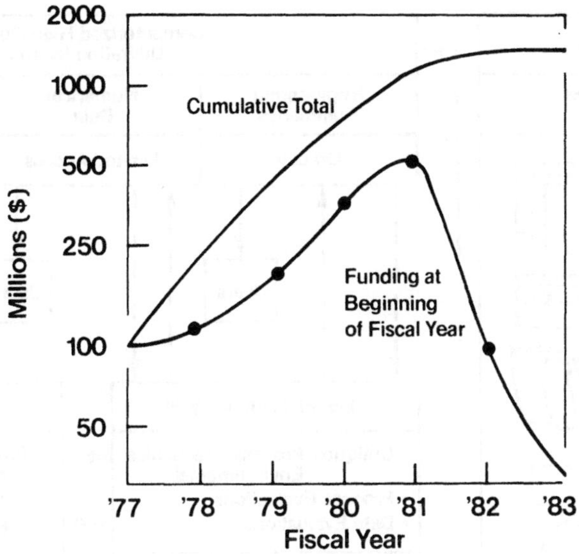

Figure 2.
PETC Technical Information Processing Procedure

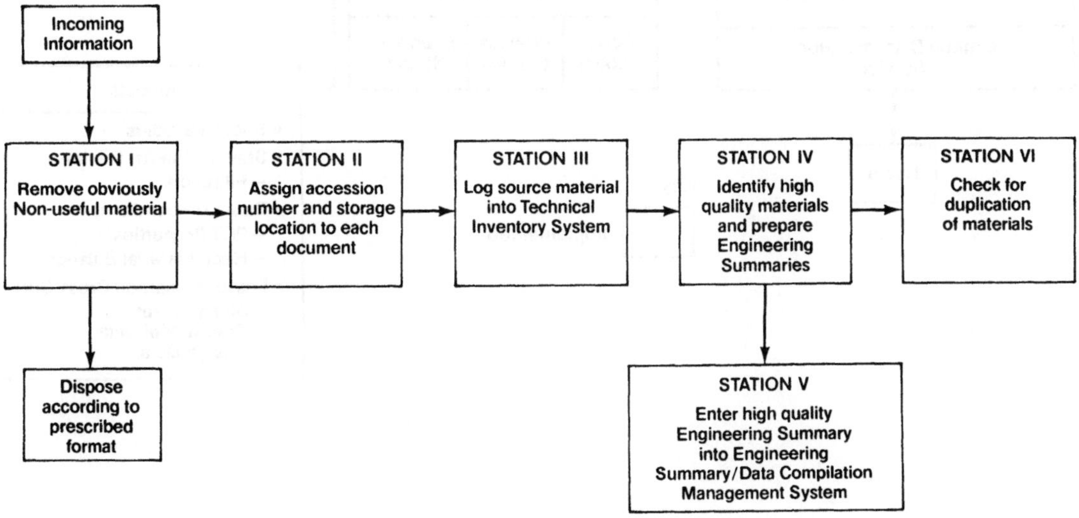

Figure 3.
Ideal LTDB System

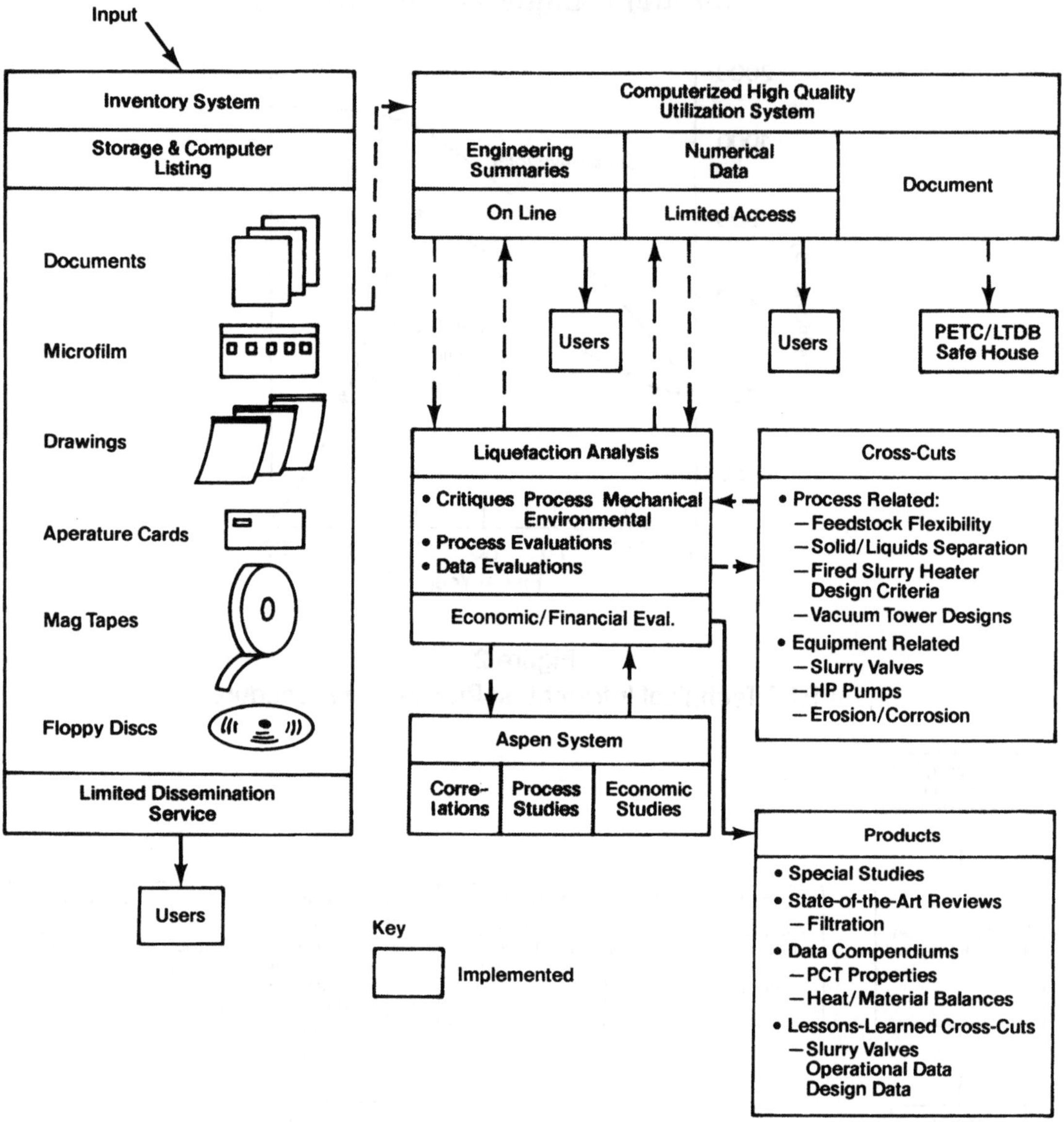

Figure 4.
Liquefaction Technology Data Base Design Schema

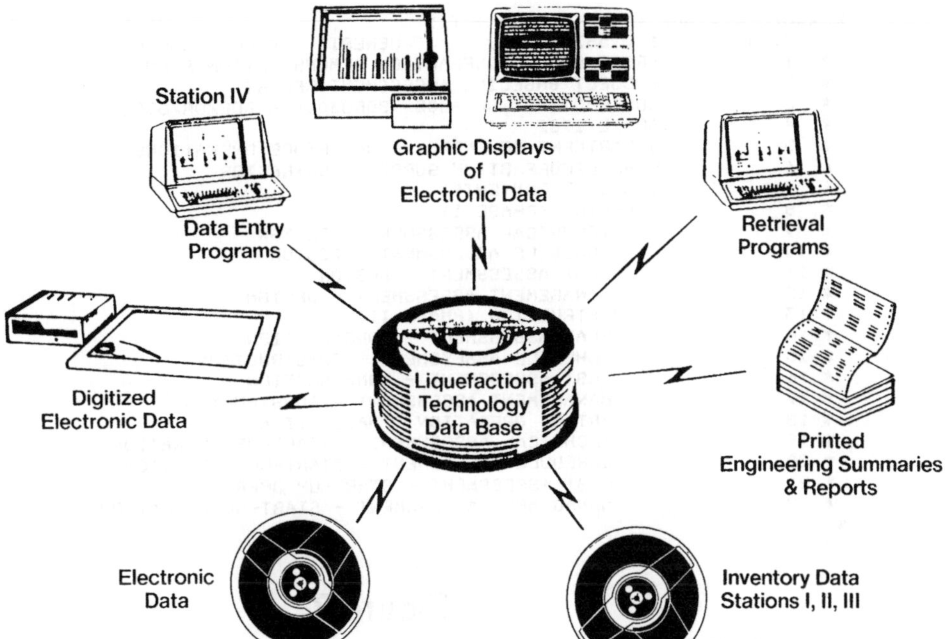

Figure 5.
LTDB Inventory Contents *
(Total, 11,153)

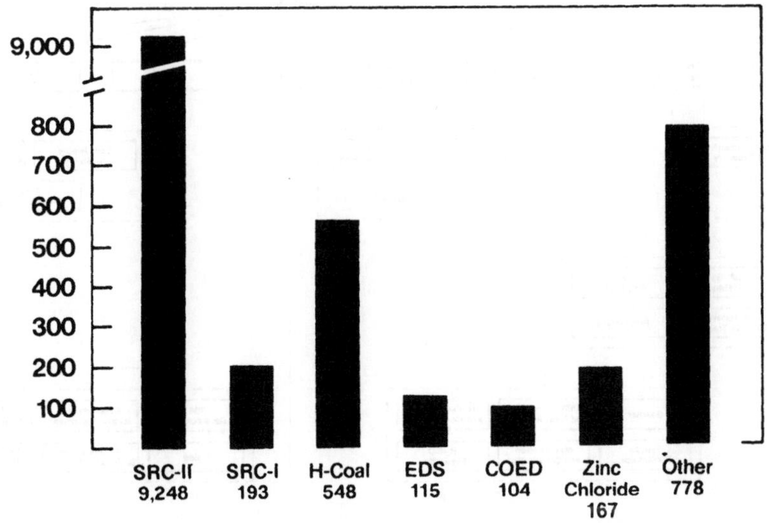

* As of June 1, 1983.

Figure 6.
Example of a String Search in the Technical Inventory System

```
keyword(s): DOE + OR + 03054 + T11
     7116           1    562   ORNL         GENERIC    4 HC PROGR N
  *  1              SRC I COAL LIQUEFACTION DEMONSTRATION PLANT
  *  2              PROJECT BASELINE ASSESSMENT REPORT
  *  3              CONTRACT NO. DE-AC05-78OR03054 - DOE/OR/03054-T11
  *  4              APRIL 1982
  *  5              DEPARTMENT OF ENERGY - OAK RIDGE OPERATIONS
  *  6              COAL LIQUEFACTION SUPPORT CONTRACTOR
  *  7              TABLE OF CONTENTS:
  *  8                DESIGN (PHASE I)
  *  9                  TECHNICAL ASSESSMENT - DESIGN
  * 10                  SCHEDULE ASSESSMENT - DESIGN
  * 11                  COST ASSESSMENT - DESIGN
  * 12                  MANAGEMENT ASSESSMENT - DESIGN
  * 13                CONSTRUCTION (PHASE II)
  * 14                  PLAN ASSESSMENT - CONSTRUCTION
  * 15                  SCHEDULE ASSESSMENT - CONSTRUCTION
  * 16                  COST ASSESSMENT - CONSTRUCTION
  * 17                  MANAGEMENT ASSESSMENT - CONSTRUCTION
  * 18                START UP OPERATION (PHASE III A AND B)
  * 19                  TECHNICAL ASSESSMENT - START-UP OPERATION
  * 20                  SCHEDULE ASSESSMENT - START-UP OPERATION
  * 21                  COST ASSESSMENT - START-UP OPERATION
  * 22                  MANAGEMENT ASSESSMENT - START-UP OPERATION
$
```

Figure 7.
LTDB Classification Structure to Plant Area Level

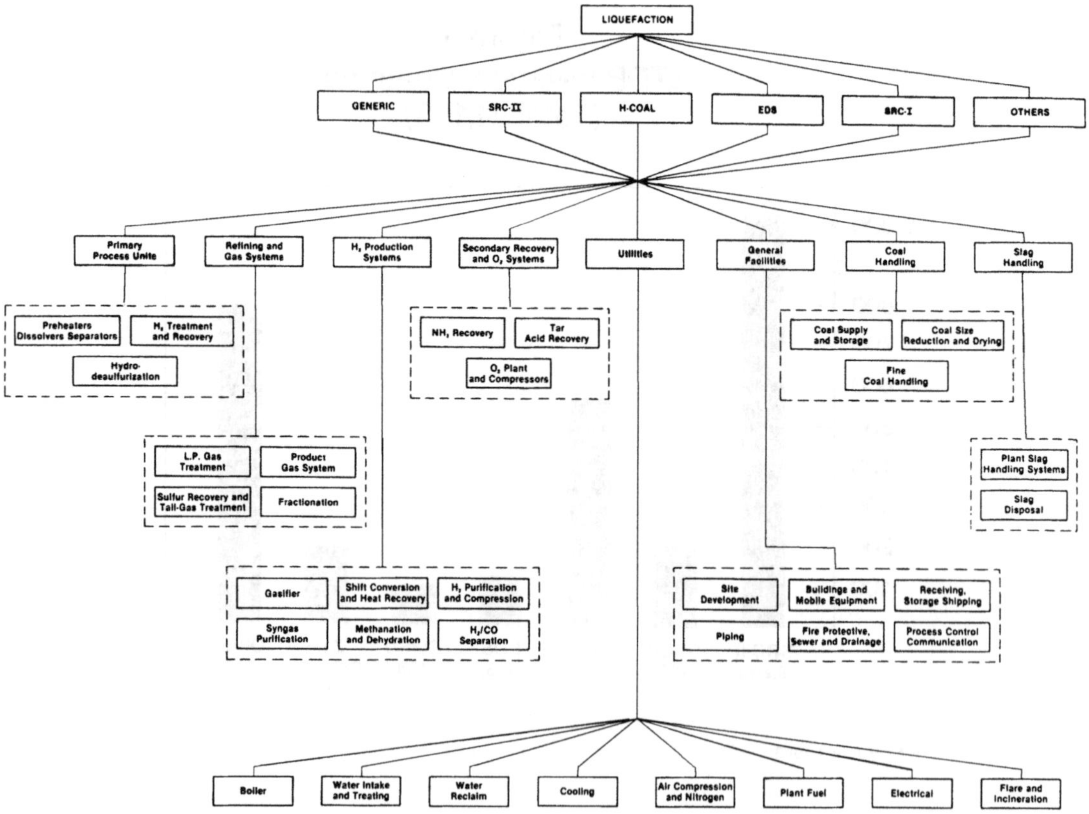

Figure 8.
LTDB Indexed Classification Structure

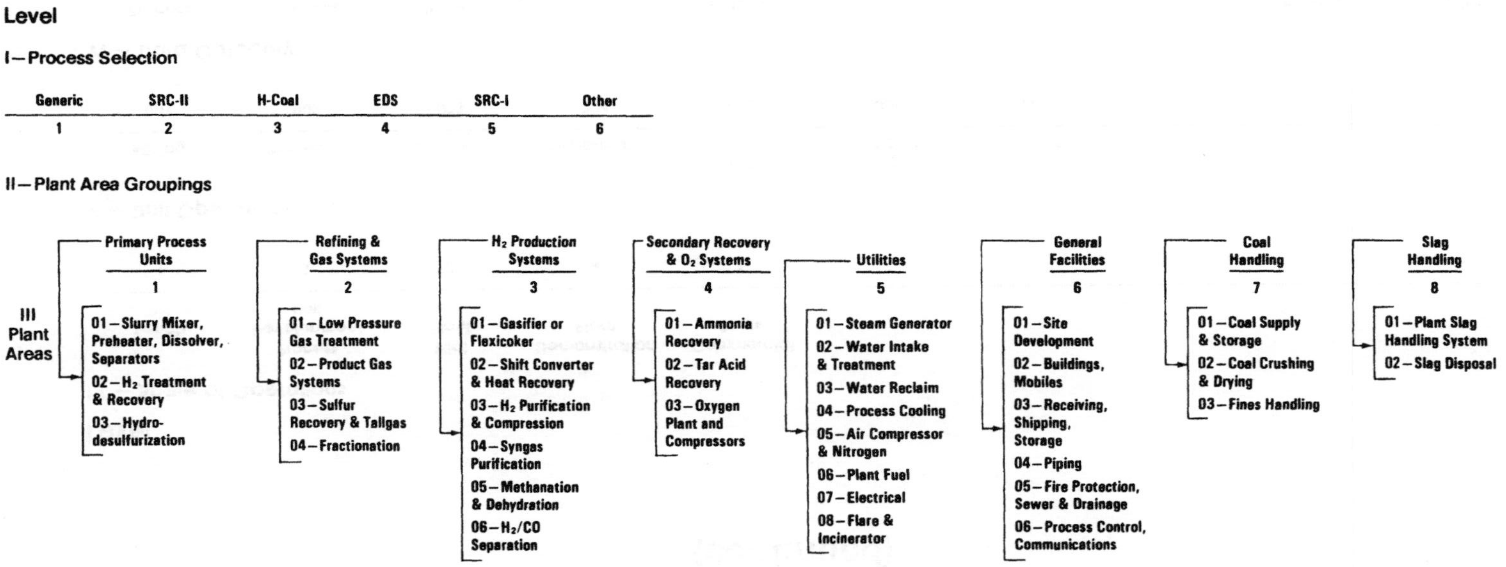

Figure 8.
LTDB Indexed Classification Structure (continued)

Level

IV — Scale of Operations

Bench	Process Development Unit	Pilot Plant	Demonstration Plant	Commercial Scale
1	2	3	4	5

V — Unit Operations

Mixing	Preheater	Dissolver	Separators	Heat Recovery	H_2 Preheater	Pumps
001	002	003	004	005	006	007

VI — Data Category

Process Design Specifications	Process Operations Specifications	Process Physicochemical Specifications	Materials Data	Process Performance Data	Environmental Data	Equipment Data	Economic Data	Project Management Data	Product Utilization Data
01	02	03	04	05	06	07	08	09	10

VII — Data Status

Raw Measurements	Reduced Data	Correlated Data	Estimates	Not Applicable
01	02	03	04	05

Figure 9.
Example of a Keyword Search in the Engineering Summary/Data Compilation Management System

```
$ LTDB
          LIQUEFACTION   TECHNOLOGY   DATA    BASE (LTDB)
                 RETRIEVAL  SYSTEM
            VERSION 0.3 (UNRELEASED 9/02/83)

         0. THESAURUS = RETURN VALID KEYWORD PHRASES CONTAINING
                        DESIRED EXPRESSIONS
         1. KEYWORD = DISPLAYS DESCRIPTIONS WHOSE ENGINEERING SUMMARIES
                      CONTAIN THE KEYWORD.
         2. ACCESS# = DISPLAYS THE ENGINEERING SUMMARY OF DOCUMENT WITH
                      ENTERED ACCESSION NUMBER.
         3. TITLE   = DISPLAYS THE TITLES CONTAINING THE ENTERED WORDS.
         4. INDEX # = DISPLAYS DESCRIPTIONS OF ALL PAPERS
                      WITH THE GIVEN ENGINEERING INDEX NUMBER
         5. AUTHOR  = DISPLAYS DESCRIPTIONS OF ALL PAPERS
                      WRITTEN BY GIVEN AUTHOR.
         6. COMPANY = DISPLAYS DESCRIPTIONS OF ALL PAPERS
                      WITH AUTHORS AFFILIATED WITH COMPANY.
         7. HELP    = SUPPLIES INFORMATION ON THE USE OF
                      THIS RETRIEVAL SYSTEM
         8. EXIT    = EXITS THIS SYSTEM
       ENTER NUMBER OF FUNCTION DESIRED AND KEYWORD ( IF NECESSARY )
           E.G.  1,NAPHTHA
       FUNCTION : 1,ENVIRONMENTAL


          KEYWORD PHRASE 'ENVIRONMENTAL' IS CURRENTLY NOT IN THE DATA BASE
          THE THESAURUS IS BEING SEARCHED FOR AN APPROXIMATE MATCH
                    PLEASE BE PATIENT
    THE COLLECTION OF KEYWORD PHRASES CONTAINING THE ENTERED PHRASE
    (I.E. THE LOGICAL INTERSECTION  OF THE ENTERED PHRASE WITH THE KEYWORD SET)
CONTAINS    5 RECORDS
      1 ENVIRONMENTAL REGULATIONS
      2 ENVIRONMENTAL IMPACT STATEMENT
      3 LEVEL ENVIRONMENTAL ASSESSMENT
      4 ENVIRONMENTAL ANALYSIS
      5 ENVIRONMENTAL IMPACT

    ENTER  1,(NUMBER) TO SEARCH DATA BASE ON THAT KEYWORD PHRASE
           2 TO RETURN TO THE MAIN MENU
           3 TO EXIT PROGRAM : 1,1
$EARCH TYPE = KEY WORD    WORKING SET =   1   0   0   0   0   0   0   0   0   0
SEARCH KEY = ENVIRONMENTAL REGULA     CURRENT SUMMARY FROM WORKING SET =   1
```

Figure 9.
Example of a Keyword Search in the Engineering Summary/Data Compilation Management System (continued)

```
==============================================================================
COMPARISON OF CAPITAL COSTS AND OPERATING CHARACTERISTICS OF BENFIELD AND
SELEXOL ACID-GAS TREATMENT PROCESSES FOR THE SRC-II DEMONSTRATION PLANT
AUTHOR : DUNCAN, JM                                ACCESSION #: 4015.1
CORPORATE AFFILIATION : SHELL DEVELOPMENT                         PAGE =   1
==============================================================================
     The Benfield and Selexol Acid Gas Treatment Processes have been considered
for use in the SRC-II Demonstration Plant.  Calculations were carried out to
determine the power requirements of the two processes.  Capital costs are
calculated for each of the treatment schemes.
------------------------------------------------------------------------------
     The power requirement for hydrogen compression in the Benfield Unit, at
the design conditions of 81,185 SCFM hydrogen flow and 1320 PSI pressure drop,
was calculated to be approximately 4.7 percent greater than that calculated
for the Selexol process.
     At the time of this study, no federal emissions standards for solvent-
            *** THERE IS MORE TEXT TO THIS ENGINEERING SUMMARY ***
==============================================================================
FUNCTIONS:   1. = VIEW ANOTHER ENGINEERING SUMMARY FROM WORKING SET (+,-N,OR #)
             2. = DISPLAY DETAIL INFORMATION ABOUT THIS PAPER
             3. = (+,-)PAGE OF SUMMARY    4. = BACK TO WORKING SET
             5. = MAIN MENU               6. = HELP    7. = EXIT  FUNCTION : 3,+
69f69fPAGE =   2
#@============================================================================
refined coal processes were in existence; thus, a comparison was made between
the projected sulfur emissions from the demonstration plant and the EPA stan-
dards for sulfur emissions from coal gasification plants. Total sulfur emissions
exceeded the EPA standards by about 11 percent.  Therefore, if the solvent-
refined coal process standards are similar to the gasification standards, addi-
tional processing of the flue gas will be required.
     This study concludes that the total capital cost of the Selexol Unit
would be $49.395 million and that of the Benfield Unit would be $56.88 million.
     A small amount of proprietary data has been deleted from the documents.
                         (69 PAGE DOCUMENT)
            *** THERE IS MORE TEXT TO THIS ENGINEERING SUMMARY ***
==============================================================================
FUNCTIONS:   1. = VIEW ANOTHER ENGINEERING SUMMARY FROM WORKING SET (+,-N,OR #)
             2. = DISPLAY DETAIL INFORMATION ABOUT THIS PAPER
             3. = (+,-)PAGE OF SUMMARY    4. = BACK TO WORKING SET
             5. = MAIN MENU               6. = HELP    7. = EXIT  FUNCTION : 3,+
69f69fPAGE =   3
#@============================================================================
                  (2 DIAGRAMS, 2 GRAPHS, 4 TABLES)




            *** THIS IS LAST PAGE OF ENGINEERING SUMMARY ***
==============================================================================
FUNCTIONS:   1. = VIEW ANOTHER ENGINEERING SUMMARY FROM WORKING SET (+,-N,OR #)
             2. = DISPLAY DETAIL INFORMATION ABOUT THIS PAPER
             3. = (+,-)PAGE OF SUMMARY    4. = BACK TO WORKING SET
             5. = MAIN MENU               6. = HELP    7. = EXIT  FUNCTION : 5
```

Figure 10.
Example of Multiple Title Listing for a Given Keyword or Keyword Combination

```
LIQUEFACTION   TECHNOLOGY   DATA    BASE (LTDB)
           RETRIEVAL  SYSTEM
   VERSION 0.3 (UNRELEASED 9/02/83)

 0. THESAURUS = RETURN VALID KEYWORD PHRASES CONTAINING
                 DESIRED EXPRESSIONS
 1. KEYWORD  = DISPLAYS DESCRIPTIONS WHOSE ENGINEERING SUMMARYS
               CONTAIN THE KEYWORD.
 2. ACCESS#  = DISPLAYS THE ENGINEERING SUMMARY OF DOCUMENT WITH
               ENTERED ACCESSION NUMBER.
 3. TITLE    = DISPLAYS THE TITLES CONTAINING THE ENTERED WORDS.
 4. INDEX #  = DISPLAYS DESCRIPTIONS OF ALL PAPERS
               WITH THE GIVEN ENGINEERING INDEX NUMBER
 5. AUTHOR   = DISPLAYS DESCRIPTIONS OF ALL PAPERS
               WRITTEN BY GIVEN AUTHOR.
 6. COMPANY  = DISPLAYS DESCRIPTIONS OF ALL PAPERS
               WITH AUTHORS AFFILIATED WITH COMPANY.
 7. HELP     = SUPPLIES INFORMATION ON THE USE OF
               THIS RETRIEVAL SYSTEM
 8. EXIT     = EXITS THIS SYSTEM
ENTER NUMBER OF FUNCTION DESIRED AND KEYWORD ( IF NECESSARY )
    E.G.  1,NAPHTHA
FUNCTION : 1,COAL/WATER SLURRY

SEARCHING ON  = KEYWORD     USING  = COAL/WATER SLURRY
 3 RECORDS LOCATED    WORKING SET =   1   2   3   0   0   0   0   0   0   0
==============================================================================
                 TITLE                                     AUTHOR      DATE
1  ALTERNATE GASIFIER SCHEMES FOR THE SRC-II DEMONSTRA    LESSIG, DP  06/30/80
2  CHOICE OF SUPPLEMENTAL FEED FOR GASIFIER IN THE SRC    SHOJI, K    06/06/80
3  TECHNICAL EVALUATION: SUPPLEMENTAL HYDROGEN SOURCE     CONNELL, A  07/07/80
$#============================================================================
 ENTER FUNCTION:  1. = DISPLAY THE ENGINEERING SUMMARY OF SELECTED DOCUMENT
                  2. = DISPLAY NEXT 10 OR LESS DESCRIPTIONS
                  3. = PUT DISPLAYED DESCRIPTIONS INTO WORKING SET
                  4. = SELECT A 'WORKING SET' OF 10 DOCUMENTS
                  5. = RETURN TO MAIN MENU    6. = HELP   7. = EXIT
     FUNCTION :7
```

Figure 11.
Relation of Models to Coal Liquefaction Data

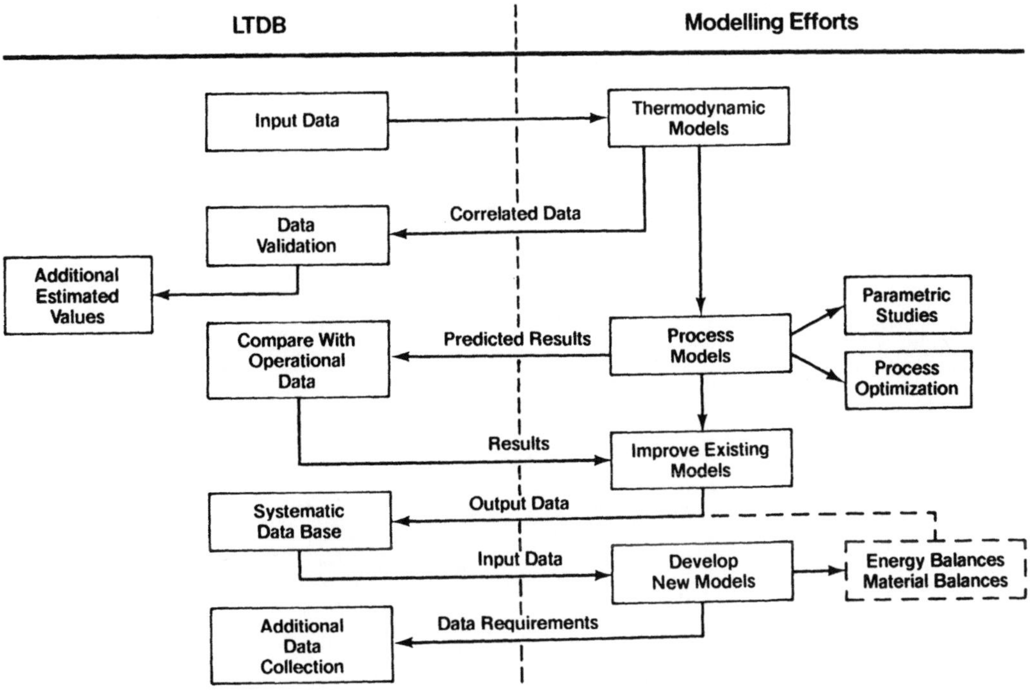

Dr. Peters received his Ph.D. in Physics from Ohio University in 1970. Since then, he has served in various R&D management assignments in the Coal Technology field with the U.S. Bureau of Mines, Energy Research and Development Administration (ERDA), and the Department of Energy (DOE).

9. ACKNOWLEDGMENTS

The authors wish to acknowledge the following individuals and organizations who have been involved in the development of the National Coal Liquefaction Technology Data Base System: William P. Haynes, Dr. Walter Fuchs, and Richard R. Desantis from PETC; David P. Mottorn, Dr. Keith Smith and Roger Duronio, from JWK International; and John S. Gordon, from the Engineering Societies Commission on Energy.

16

PROBLEMS IN INTEGRATING AND AUTOMATING A MULTI SITE/MULTI SPONSOR TEST DATA COLLECTION AND REPORTING SYSTEM

Patrick A. Race
Department of theArmy
United States Army Tank-Automotive Command
Warren, Michigan 48090

Abstract

This paper attempts to describe the problems that are being encountered as two US Army Commands attempt to upgrade existing systems for reporting the data captured during tests on tanks and automotive vehicles. The combination of the organizational structure and the variety of data processing systems that must be combined presents an opportunity to examine an automation technology problem of complexity. The problems encountered are example of both the specific and the general. The specific problems arise from the need to fashion a system to certain requirements and utilizing for the most part existing resources. The general problems that are institutional, personnel and technical are examined in the context of the US Army's test data collection systems.

1. INTRODUCTION

1.1 REPORT STRUCTURE

This report presents information about an automation project in a way that focuses attention on the problems that exist in any large automation project which attempts to integrate many disparate elements. Considerable information is provided on existing Army systems for test data collection and a comprehensive description of the model of a desired integrated system that will be called the Integrated Test Data Collection System (ITDCS). In the main section of the report those problems seen as impeding the implementation of the Integrated System are addressed.

1.2 BACKGROUND

The US Army maintains many facilities to prepare for and achieve its various missions. The administration of materials needed to complete these missions is carried out by specialized commands. The command that manages the development, testing, acquisition and fielding of all Army equipment is called the Development and Readiness Command (DARCOM). Within DARCOM the development and support of tanks and automotive equipment is accomplished by TACOM (Tank Automotive Command). Testing functions are administered by TECOM (Test and Evaluation Command). These two commands are in turn divided into many directorates, offices or divisions that perform the actual developmental or testing functions.

The majority of the tests are scheduled by Engineering, Project Managers and Product Assurance organizations at TACOM. The office which schedules the test is called the "test sponsor". Testing is performed in order to determine logistic support, maintenance times, maintenance echelons, performance, reliability/durability, and the quality of an item.

TECOM administers many test sites designed for particular environments and applications. The test site organizations collect data according to the requirements specified in a test plan. Usually a data report is written whenever any maintenance is required. This report is stored either in a manual or electronic file and transmitted to the test sponsor.

A test incident report provides maintenance, environment, vehicle and component information along with the action taken to resolve the problem that caused the incident. Many of the data elements are coded to facilitate retrieval procedures.

2. EXISTING RESOURCES

Test data collection and storage became automated as independent data processing systems within TACOM and TECOM were developed. As the use of these sytems increased, problems with the availability and accuracy of data became evident and many test data automation issues were raised. These problems and issues stem from the following sources.

(1) Lack of compatability of computer and data communication resources between TECOM test sites and TACOM organizations.
(2) Different interpretations of requirements for data collection at test sites and by test sponsors.
(3) Experimental projects with computer systems that gave test data users a greater interest in automated systems and increased their expectations for the existing resources.

In addition to the problems experienced in handling the data from tests, the increasing availability of computer equipment and the increasing desire for additional automation procedures provided additional impetus for a complete restructuring of the way test data is processed at test sites and used by test sponsors.

2.1 TACOM RESOURCES

The computer resources in use at TACOM at the initiation of the process to create ITDCS are listed below:

(1) Numerous terminals and microcomputers running software that provided interactive capabilities.
(2) Access to several large timesharing mainframe computers through commercial networks.

Test information was stored in data base management systems and data manipulation was accomplished through the use of standard computer languages.

2.2 TECOM RESOURCES

The computer resources for TECOM exist at individual test sites and at TECOM Headquarters. There are varying degrees of computer utilization at the test sites. One of the more advanced test sites uses a microcomputer as a front end to its mainframes and they have developed a variety of programs to assist in managing data. Access by TACOM to the test data is available through timesharing. In addition to the computer equipment used for test data, other equipment and software packages are availabe

Other test sites use manual systems to collect, store and manipulate data although computer resources are available at those sites or through time sharing.

3. ADDITIONAL RESOURCES

There exists within the many Army Commands a large number of computer resources that may be utilized on a time sharing basis by other commands. There are also continuing programs to upgrade existing systems and to create new capacities and resources. The following paragraphs describe additional resources that can be made available for the ITDCS.

3.1 COMMUNICATION NETWORKS

In 1969 an Defense Advanced Research Projects Agency (DARPA) project yielded the first packet-switching network which evolved into a defense communications network. This communication network was created to give high speed, secure data transmission capabilities. Host systems are connected to this communications network through Host Front End Processors (HFEP), Terminal Emulation Processors (TEP) or directly. Links to this communications system will be installed at all TECOM test sites and at various offices within TACOM.

3.2 DIRECTORATE OFFICE AUTOMATION EQUIPMENT

Present office automation studies are recommending generic types of equipment for the automation of many functions that are performed in TACOM and TECOM organizations. The recommendations generally specify supermicro computers that use the UNIX operating system and have several terminals attached. The presence of these in the TACOM Product Assurance Directorate and other test sponsor's offices will allow the test sponsors the capability of electronic manipulation of the data relevant to their needs.

4. DESCRIPTION OF THE INTEGRATED TEST DATA COLLECTION SYSTEM (ITDCS)

4.1 SYSTEM FUNCTIONS

A replacement for the various existing systems that collect, store and provide outputs of test data has long been sought by those organization that use the data from TECOM tests and by the TECOM organizations that conduct the tests. The Integrated Test Data Collection System that this paper assumes performs six functions. They are listed below:

(1) Data Entry to Disk
(2) Maintain Test Site DBMS
(3) Maintain Test Sponsor Access File
(4) Connection with Government Communication System
(5) Maintain Test Sponsor DBMS
(6) Maintain Inter Command Access Files

In order to relate technical and institutional problems in implementing the ITDCS it is necessary to give an administrative description in addition to the functional one given above. The elements of the administrative description are as follows:

(1) Test Site Data Management System - this includes the Data Entry, Test Site DBMS, Access File and Network Connection functions.
(2) Site/Sponsor Communication System - this is the Government Communications Systems.
(3) Test Sponsor Data Management System - this system includes the Test Sponsor DBMS, Access Files and Network connection function.

The implications of institutional relationships that necessitate an administrative description will be considered in the section that describes the Problems in Implementing the ITDCS.

4.2 FUNCTIONAL DESCRIPTION

The Integrated Test Data Collection System (ITDCS) comprised of the above listed six elements, includes functions that are presently in the existing systems and it has functional parts that are not presently used. Some of the additions will come from currently existing resources that were not used by the data collection organizations and other additions will be the result of acquisitions made for the specific purpose of implementing the ITDCS. A detailed discussion of each function follows:

Data Entry

The data entry capacity that is needed for entering data provided by test data collectors should be provided by a module of the DBMS discussed below. There needs to be multi-station entry and an editing capability.

Maintaining DBMS at Test Site

The computer that provides for data entry should also be able to support a DBMS that is capable of sophisticated manipulation of the data. Screen formatting, sorting, transaction processing, computation and restructuring of databases are all functions that the DBMS should provide.

Maintaining Test Sponsor Access File

The operating system, DBMS or another file processing program should allow the creation of files in standard format so that specific subsets of the test data can be transmitted to test sponsors or others who have a need to use it. These files should be stored in a manner that allows potential user to access them with a minimum of difficulty.

Connection with Government System

This element should provide high speed data transmission capabilities to users who are connected to its nodes. Both the test sites and the test sponsors can utilize this network for test and other data transmission.

Maintaining Test Sponsor DBMS

The computer system that supports the DBMS for TACOM must allow for multi-user access and have the same capabilities as the one for the test sites.

Maintaining Inter Command Access File

The operating system, DBMS or another program must be provided to allow access to files of information by any of the test sponsors or other users of the DBMS.

4.3 ADMINISTRATIVE DESCRIPTION

The three entities that are described in the administrative description of the ITDCS are distinct elements in the organization of the Army. The Test Site Data Management System will exist within the Data Collection and Test Management organizations at the test sites. The communications function is provided by a government communications network that is under the Defense Communication Agency. The Test Sponsor Data Managing System will exist within a designated office at TACOM. Attention must be given to these institutional relationships because the data processing resources for ITDCS are constrained by the policies of the commands they belong to. In other words the policies governing availability of equipment, addition of staff, etc., for the Test Sponsor Data Management System at TACOM are not necessarily the same as those which TECOM requires the test sites to observe. The three elements are discussed below with an attempt to describe specifically what data processing resources are needed for each.

Test Site Data Management System

There are four components of the system that are needed at each test site. They are:

(1) Data Entry to Disk
(2) Maintaining a DBMS
(3) Maintaining Files for Test Sponsor Reading
(4) Connection to the communications network

The computer system at each test site must have the capacity to run data entry, DBMS, file transfer and communication programs. It must be connected to a Host computer having high speed data transmission capability and a storage capacity that will accommodate a complete test. Support personnel must have expertise in operating the computers and peripheral equipment and be able to maintain all of the listed functions.

Test Site/Test Sponsor Electronic Communication System

The test sites and TACOM will need to develop the expertise to implement and maintain connections from host computers or TEPs to the communication network nodes.

Test Sponsor Data Management System

There are three components of the Test Sponsor Data Management System that are needed at TACOM. They are:

(1) Connection to the communications network
(2) DBMS for Test Sponsors
(3) Maintaining Files for Other Commands.

The computer system at TACOM must have the capacity to run communication, DBMS and file transfer programs. There must be high speed communication and the ability to store all data from an individual test. There needs to be a sufficient number of terminals available to the test sponsors so that data can be accessable. Support staff will be required to train users and maintain the hardware and software.

4.4 SUMMARY OF THE ITDCS

The ITDCS will attempt to provide data from various TECOM test sites to test sponsors at TACOM through the use of front end DBMS at the test sites, files transferred through the communication facilities and a DBMS within TACOM that is accessable by the individual test sponsors.

5. PLAN FOR DEVELOPING AN INTEGRATED TEST DATA COLLECTION SYSTEM

The plan for developing the ITDCS will be to take each of six functions of the ITDCS and assess to what degree the current and available resources can provide the needed function. From this assessment will come the action appropriate to each option. Listed below are the actions and a short description of each.

(1) USE CURRENT RESOURCE - The existing resource, as is, or with slight modification will fill the need.
(2) USE AVAILABLE RESOURCE - The need can be filled by a currently available resource that was not used in test data collection before.
(3) USE UPGRADED RESOURCE - The need can be filled by anticipated addition or upgrade of existing equipment.

TABLE 1

FUNCTION	OPTION
Data Entry to Disk-Test Site Number1 (TS1)	USE CURRENT
Data Entry to Disk-Test Site Number 2 (TS2)	USE UPGRADE
Test Site DBMS-TS1	USE AVAILABLE
Test Site DBMS-TS2	USE UPGRADE
Test Site File Creation and Storage-TS1	USE CURRENT
Test Site File Creation and Storage-TS2	USE UPGRADE
Connection of TS1 to Communication Network	USE UPGRADE
Connection of TS2 to Communication Network	ACQUIRE UNIQUE
Electronic Communication Facility	USE UPGRADE
Connect Test Sponsors Communication Network	USE UPGRADE
Test Sponsor DBMS-TACOM	USE UPGRADE
Test Sponsor File Creation and Storage-TACOM	USE UPGRADE

(4) ACQUIRE UNIQUE RESOURCE - The need can only be filled by the acquisition of additional resources.

As each function is assessed the correct option will be selected. The result of this process will be the elements of a plan for implementing the ITDCS. The following table shows which action is applicable to each function of the ITDCS. TECOM has many test sites, but for purposes of space and clarity only two test sites will be included in Table 1.

The above listed actions constitute a generalized plan for assembling an ITDCS from the various resources that are available to the TACOM and TECOM organization charged with data handling functions.

The problems specific to the successful completion of each action category will be discussed in the next section.

6. PROBLEMS IN IMPLEMENTING THE PLAN FOR AN ITDCS

The problems that are encountered in trying to implement the plan for transition from the use of existing resources to full utilization of the Integrated Test Data Collection System are of two types. Some of the problems can be considered to be of a general nature, while others are specific to this system. Each type will be discussed in detail.

6.1 SPECIFIC PROBLEMS

The problems specific to the implementation of ITDCS in the various TACOM and TECOM organizations are listed below.

(1) Selection and installation of data entry, DBMS, and file storage software for test sites without these capabilities
(2) Creation and installation of communication equipment between Test Sites/ Test Sponsors and the communication network
(3) Acquisition of communication network documentation
(4) Installation of office automation equipment in TACOM
(5) Selection and installation of DBMS at TACOM
(6) Installation of file storage software on TACOM Computers.

6.2 GENERAL PROBLEMS

The problems of a general nature can be divided into at least three different types. Those problems of the first type are of a institutional nature and are the most numerous. The second type of problem is associated with the personnel available to support the system and the third type of problem is technical.

6.2.1 Institutional Problems

The institutional problems associated with implementing ITDCS are of four major types. They are listed below along with a definition for each.

Fiscal-funding is unavailable for providing the function. This type of problem can prevent a

TABLE 2

ACTION	INSTITUTIONAL PROBLEMS
USE CURRENT	INFORMATION DISSEMINATION
USE AVAILABLE	Above and INFORMATION ACQUISITION
USE UPGRADE	Above and TIME LAGS
ACQUIRE UNIQUE	Above and FISCAL

Function from being implemented. In the ITDCS this problem plays no roll because funds will be committed for each function.

Time lag - the resource will not be available until a specified time in the future. This is actually how fiscal considerations affect the ITDCS since future procurements and the length of the procurement cycles often delay the implementation of a system function.

Information acquisition - needed information is not available to users of the system due to circumstances outside the system. This problem prevents utilization of a system function even though all the equipment is in place.

Information dissemination - needed information is unavailable due to incomplete internal documentation procedures. This problem can be a difficult one to overcome in a large decentralized organization.

These institutional problems can be associated in a systematic way with the action categories defined in Section 5.

6.2.2 Personnel Problems

While it is true that the lack of staff can always be classified as a fiscal problem as defined above, such classification helps little in an attempt to identify problems associated with the implementation of complex data processing systems. This section will identify two types of problems that exist because of less than adequate staffing in the various organizations concerned with test data.

The Under Staffing Problem

This problem has to do with the situation where individuals have the knowledge and skill to determine and perform the required action but there is not sufficient time in their regular schedule of work to perform the activities needed.

The Lack of Training Problem

This problem has to do with the situation where the individual has the skills needed to perform the activity but needs additional training on the specific equipment or program that supports the system function.

Both types of problems exist in the various TACOM and TECOM organizations and different solutions are required for each.

6.2.3 Technical Problems

A technical problem is one whose solution requires a level of effort and knowledge not normally associated with the personnel who staff positions of data processing support for the test data collection systems. Problems requiring technical solutions are those that will be solved by a special contractual addition of resources for the sole purpose of solving the problem. In implementing the plan for ITDCS there are two functions that contain technical problems as defined above. These both are concerned with the connection of host computers to the communication network and the solution to each will be provided by a technical expertise secured through special contract.

7. CONCLUSION

The implementation of a system like ITDCS will take place over a period of several years. The solutions to the problems described above involve many activities and require the integration of several systems. The specific problems discussed above will be solved as the Army procures additional equipment and implements the software to run on this new equipment. Of the general problems discussed the Institutional Problems will be of continuing interest to those who are responsible for integrating complex systems that cut across the formal boundaries of large government and corporate organizations. Methologies for timely provision of essential information to widely dispersed centers require that on-line documentation be provided as an integral part of any system similar to ITDCS.

8. BIOGRAPHY

Mr. Race is a General Engineer at the Tank-Automotive Command at Warren, Michigan. He is currently working in the RAM Data Branch of the Product Assurance Directorate. Recently, he was Co-Chairman of the Test Data Collection Study Group which is a combined effort of TACOM and TECOM to correct current problems and prepare for future needs of tank automotive test data users. He has a BSE in IE from the University of Michigan.

17

OPTICAL STORAGE OF DIGITAL DATA ON VIDEO DISCS FOR DISTRIBUTING ENGINEERING DATA AND TECHNICAL DOCUMENTATION

Jane N. Ryland
Reference Technology, Inc.
Boulder, Colorado 80301 U.S.A.
(303) 449-4157

ABSTRACT

Many of the optical storage products currently announced or expected in the next few years, including both write-once and erasable implementations, offer attractive solutions for capturing engineering data or technical documentation as it is created. There are additional needs, however, which are not satisfied by any of these approaches, such as distribution of the information to the users who need to work with it.

A newly emerging approach, being pursued by Reference Technology, Inc., uses standard video discs to store large volumes of digital information and also solve the problem of mass distribution. This paper discusses the particular benefits which this approach offers for distributed applications, as well as the challenges which must be overcome to make this approach viable.

The benefits include:

- o high media capabilities (in excess of 2 GB)
- o easily replicable media at very low cost
- o use of low-cost, high-performance, read-only hardware
- o stable, proven media

The challenges to be overcome include:

- o error correction to achieve computer standards for bit error rates
- o attainment of drive performance compatible with large data volumes and suitable for multi-user service
- o computer systems interfaces

INTRODUCTION

Optical storage products have been the subject of considerable publicity in recent years. Several write-once products have been announced, and more announcements are expected before the end of the year. Disclosures of advances in erasable optical technologies have generated considerable interest. While these products offer attractive solutions for capturing engineering data or technical documentation as it is created, there are additional needs that are not satisfied by any of these approaches, such as distribution of the information to the users who need to work with it.

Large volumes of engineering data and technical documentation, including digitized images, pose problems if only a single copy is created and attached to a central computer and then made available via telecommunications links. The volume of data that made storage difficult until the advent of optical products' greater density and reduced costs easily choked a telecommunications link from the standpoint of both time and cost. Transmission of 4 gigabytes, for example, would take more than a month, 24 hours a day, at 9600 baud.

Video disc technology, on the other hand, offers advantages for distribution because thousands of discs can be replicated quickly and inexpensively from a master. "Publication" of engineering data and technical documentation in this form allows this material to be searched, displayed, and analyzed by professionals at individual, microcomputer-based work stations ranging from an IBM Personal Computer through 68000-processor-based systems. In addition to being inexpensive to reproduce, the video disc is convenient to access, store, and physically transport. Information can be stored as video images by using direct frequency modulation to produce the television format commonly recorded on video discs today; but the video disc's real potential for data storage comes from a unique approach that allows the encoding of digital information in the video signal.

FROM TECHNOLOGY TO PRODUCT

The elements necessary both to demonstrate the potential of the optical video disc for digital information distribution and to make this technology a profitable business are:

- A premastering process for encoding digital information into a standard television signal so that it can be stored on an optical video disc using currently available production techniques, and permitting a combination of digital and analog data when desirable.
- A method of error detection and correction that accommodates the optical video discs, which are molded storage media on which defects can be introduced during reproduction. The method used must ensure information retrieval at a consistently high level of accuracy considered acceptable in the computer industry.
- A sophisticated drive for reading the optical video disc that meets computer industry standards for reliability, access time, and on-line capabilities. The drive should be manufacturable using the competitive production techniques that will permit volume manufacturing.
- A high-speed standard interface for computer attachment.

This combination of capabilities has been developed by Reference Technology and will offer a unique solution for distribution of virtually any kind of reference material, including engineering data and technical documentation.

MARKET TRENDS

The market demand for information stored in electronic form has increased significantly during the last decade, and virtually all studies indicate that this trend will continue. Several trends in the market are particularly favorable to the introduction of products that facilitate distribution of digital information on optical video discs. For example, there has been tremendous growth in the business of providing online database services. A study by Link Resources shows that $1 billion in revenues were generated in 1982 from providing public information in electronic form, a substantial percentage of which was from online database services, and these revenues are expected to grow to nearly $3 billion in 1987. Since a significant proportion of this information is static and therefore well suited to alternate means of electronic distribution, Link estimates that by 1990 in the United States alone, revenues from publishing data bases on video discs, not including any sales of hardware, will be nearly $0.5 billion.

Where previously information had to be converted to machine-readable form, more and more information is being directly generated by computerized or electronic collection processes. Satellites, sensors, and scanners are dramatically increasing the amount of data that must be stored. Engineering data are often the result of CAD/CAM/CAE systems, and technical documentation is often generated by computerized text generation systems.

A second major trend of primary importance to the long-term market for the optical video discs has been the growth in sales of microcomputers, particularly for business use. Microcomputers and professional work stations are rapidly reaching the point where they will dominate the market serving business- and professional-computing needs. Mainframes and minicomputers will undoubtedly remain important for some applications, but they will represent a decreasing share of this market. Large and expensive data storage systems are not suited for this environment of distributed computing power, but optical video discs will fill an important market niche by allowing large, stable data bases to be recorded and readily distributed for direct access on individual, professional work stations, while lower capacity, more expensive magnetic media will continue to be used for working space and more volatile data.

These two trends alone would confirm the value of providing information locally to these professional users who now have distributed computing capability at their fingertips; but a third major market trend suggests a specific direction in matching information storage to the professional work station. This is the growing availability of local area networks (LANs), which support communication among local users, provide the communications links to outside services, and can facilitate local sharing of such information resources as optical storage systems. A recognition of this trend dictates a product with high performance characteristics, capable of allowing queries from multiple work stations, rather than a product that can only support dedicated attachment to a single work station.

OTHER OPTICAL TECHNOLOGIES

Other optical technologies are often confused with optical video discs. Optical direct-read-after-write (DRAW) systems have received considerable attention and bear some similarities to the optical video discs. Both use optical technology, but there are important differences that dictate very different long-term markets for the two types of systems. DRAW media, whether erasable or not, are not replicable except by completely reading a disk and writing it to another machine. DRAW products, unlike optical video disc drives, allow writing online. These characteristics will put the DRAW systems more into competition with magnetic disk drives than with the optical video disc, especially when erasable DRAW products come to the market. Aside from the advantage of fast and inexpensive replication, optical video discs offer considerably lower drive costs as well as protection from destroying the information of the disc due to the read-only nature of the drive. In addition, these media have a proven record of stability of more than ten years, and is considerably less sensitive to degradation and handling than optical DRAW media.

Interactive, programmable video disc players, another version of optical technology that can use the optical video disc, allow the user to select and use the video images desired from the material on the disc by giving simple microprocessor commands. Primarily used in entertainment and training applications, these systems have low costs, ranging from $1,500 to $2,500 (exclusive of any computer or computer interface). They are, however, characterized by relatively slow access times (two to ten seconds), slow data rates, and high error rates, and are limited to a capacity of only several thousand video images.

The product approach being pursued by Reference Technology offers a unique combination of characteristics that are best suited for distributing large volumes of digital information. Like optical DRAW products, the Optical Reference System will have a high capacity for digital information, fast access times and data transfer rates, and high data integrity. Like the optical video disc, the Optical Reference System will use inexpensive, easily replicable media, and offer low drive costs.

OPTICAL REFERENCE SYSTEM APPLICATIONS

The Optical Reference System approach has wide applicability for distributing information of virtually any type, but especially that which can be enhanced by the use of a personal computer or work station. Publishers of publicly available data bases see this as having the potential to dramatically augment online services by providing a more cost-effective method of searching and analyzing these data bases. With the proliferation of computer use for typesetting and publishing, more and more published information is already in machine-readable form, facilitating conversion to an optical video disc as the published medium.

While this information publishing market will probably take some time to develop, there are many near-term applications for distribution of information within a single organization or within a family of information users. Engineering data and technical documentation are excellent examples of this application. These applications require large storage capacities in a small form factor, especially when digitized images are a significant component of the information. Fast direct access to any piece of information is a specific requirement, as is the ability to rapidly transfer data from the storage device to the host processor. High integrity of the information is critical as well. Finally, the ability to create many copies of the same data set quickly and at very low cost is important in distributing the information to users, especially those who are broadly dispersed geographically. This form of distribution is much more cost-effective than trying to provide this information in a single central data processing installation with links to the remote users via telecommunications. Local storage of the information, as previously discussed, also allows the users to take advantage of the processing capabilities of their personal-computer-based work stations. One of the most exciting features made possible by the use of the optical video disc is the mixed media data base, interleaving text, video images, high-resolution digital images, and audio enhancements.

All of these characteristics combined represent a superior approach for storing and distributing engineering data and technical documentation, an approach that is consistent with the market trends in distributed processing and local area networking.

BIOGRAPHY

Jane N. Ryland, Manager of Market Support for Reference Technology, Inc., was previously Manager of Optical Systems Planning for Storage Technology Corporation. Her career in information systems has spanned 17 years, including positions as a Systems Engineer for IBM, Data Base Administrator for Virginia Polytechnic Institute and State University Computer Center Director for the Virginia Community College System, and director of a federally funded office to promote coordination in data collection for federal and state higher education agencies. Educational achievements include a BA in English and an MPA in Management Systems.

18

AN INTEGRATIVE APPROACH TO ENGINEERING DATA AND AUTOMATIC PROJECT COORDINATION

Richard L. Segal
California Institute of Technology
Pasadena, California

Abstract

A new approach to integration of automation environments is necessary in order to overcome some fundamental limitations of current methods. The most common request for integration comes when an organization finds itself with many design systems and representations most of which cannot communicate directly. Solutions to this problem have required either a proliferation of translations between systems, or the provision of a central data base with which all applications must communicate. Another aspect of integration is the provision of mechanisms which keep track of the status of design. The working-release method along with version control techniques is well known but does not directly address the compatability issue and provides only a partial solution to status keeping. Another approach to integration is to exploit the underlying structure of an engineering project. This method can result in improvement in both realms of integration described above. The potential results of such an approach include the following, each of which is described in this paper. Firstly, this approach provides a new mechanism for providing design team independence and a better definition of design team responsibility. Secondly, the concept of inter-verification is introduced as a way of providing a continuum of representations which can be kept consistent through the project structure. Thirdly, automatic conflict detection and coordination mechanisms can be implemented, eliminating much of the delay usually associated with design changes. The final portion of this paper indicates some of the other areas influenced by the use of project structure in the computational environment. This includes subdivision of verification processes, documentation systems, external object integration and automatic design systems.

1. ENVIRONMENT

An engineering organization contains an administrative structure for the purpose of managing and assigning resources, and a project structure associated with each ongoing product design. Engineering data is input, analyzed, output and communicated within and through the project structure. The project structure consists of "modules" each of which may contain "parts" which are instances of other modules. Since various modules may be in the process of design simultaneously, it is often necessary to make assumptions regarding the final form of parts of a module. These assumptions are called "requirements." The "design" of a module consists of manipulation and composition of parts. The task of a module design team is to provide design which will meet a particular set of "specifications." As specifications of a module become firmed up, the requirements placed on that module can change to reflect this process. Both requirements and specifications are constantly changing, first as the initial design team responsibilities are being worked out, and later as the details of the various designs take shape.

2. INTEGRITY

This project structure exists in every large engineering project in one form or another. By using this structure to organize the subdivision of data bases, we are suggesting a new approach to data base organization. Rather than a simple working-release set of data bases, each data base contains precisely the current representation of its design. Design team independence is accomplished through the RSD or Requirement, Specification, Design method. A module team performs its task by using requirements to represent actual parts. Specifications, though similar in form to requirements, are announced independently by each module team. Specifications are verified in terms of module design and requirements are verified in terms of part module specifications. Mechanisms to allow project refinement and keep track of the status of engineering data can be based on this structure. The working-release method of data base separation achieves independence, but does not maintain up to date interrelationships and does not guarantee the compatability of different working data sets.

We use the term verification to represent all forms of analysis, calculation, comparison and simulation procedures. Verification of specifications through whatever technique desired and verification of design correctness, which includes ordinary design system feedback and error detection, all takes place within a module. We call this "intra-verification." "Inter-verification" is the checking of correspondence between the requirements placed on a module and the actual specifications announced within that module. Any number of user modules may place requirements on a particular module. Each such requirement must be verified. By providing inter-verification for every kind of requirement, the current status of the complete engineering project is accessible at all times.

3. AUTOMATIC COORDINATION

Every kind of verification has a particular set of inputs which it can use. For example, specifications include statements like "the mass is between 50 and 60 kilograms" as well as "the circuit behavior is described by xxx" where xxx is a state machine description, or finally, "there is a socket at position 10.5, 150,88", perhaps specified through a combination of graphics and text. Clearly, a specification may use any other element of module data in its verification. Verification of design, which includes operations like design rule checking and the detection of illegal manipulations at interactive design time, may only use information in the requirements, and information in other design elements. The latter is necessary when a composition of parts, each of which may themselves be rotated or otherwise manipulated or parameterized, is performed. Suffice it to say, that a complete and formal definition of the available input sets for each form of verification can be determined.

Each verification actually uses only some subset of its available input. This set is called the constituent set of a verification. If the constituent set of every verification is known, then it becomes possible to perform automatic design coordination. There are two aspects to this process. "Verification tracking" is the task of maintaining the current status of every data item in a project. This means that every time a data item is altered, each verification containing this item in its constituent set must be invalidated or reverification must occur. This set of verifications is called the "track" of a data element. The second aspect of project coordination is called "notification." This is the task of informing designers when a change having a bearing on their design has been performed.

3.1 VERIFICATION TRACKING

A simple mechanism can be used to determine if an item is in the track of a particular design change. A procedure, called a "tracker", is provided for every kind of verification. A tracker is a surprisingly simple program, usually only a few lines of code, which just checks the type of the item which was altered and reports either positively or negatively that its associated verification has the altered item in its constituent set. For example, let us say there is a design item called "conn" which is used to join two objects together. There is a verification called "vcon" used to check the compatability of objects in a conn. A tracker for vcon, call it "tcon", would operate as follows. The parameters to a tracker include the item containing the verification to be considered and the item which is being tracked. Tcon would first read the names of the objects being joined in the conn. If the altered item is a connector type requirement on one of those two modules, then tcon returns a positive report, otherwise it returns negatively.

Trackers are all short similar programs usually far less complicated than the verification with which they are associated. However, if a verification is quite simple, rather than performing the usual check, the tracker can simply call the verification directly. This requires a third and fourth kind of response to be permitted from trackers. They are, that the tracker called the verification, which, in turn, either succeeded or failed.

The entire process of verification tracking consists of two steps. First, all the trackers which have the altered item in their available input set are called. Second, all the verifications for which a positive report is received (ie. that the item is in the track and was not reverified) are invalidated. Verification tracking should be an invisible operation which occurs whenever data is altered. It should be capable of dealing with a number of data items at one time, and should make efficient use of networks, requesting all information needed from each module at one time.

3.2 NOTIFICATION

There are two aspects to notification. The first is negotiation notification. This occurs when a specification or a requirement is changed. When a module changes an element of its specifications, all users of that module are notified. When a module changes the requirements it places on a part module, the designers of that part module are notified.

The second kind of notification is validity notification. Any time a verification tracking operation occurs, each of the designers responsible for data in the track of the changed element must be notified of the action. They are told exactly which verifications have been invalidated or reverified.

4. CONFLICT RESOLUTION

The detection and resolution of design conflicts is one of the most important tasks of an engineering project. Typically, if a design change in a low level module inherently conflicts with a high level management constraint, the problem might not be noticed for weeks or months. Usually there are regular project meetings at which the offending design data might be reported. Later, managers meet with summaries of each of their groups activities. There is no guarantee that the data in question will be reported at either of these meetings. Eventually the data will reach the party responsible for performing the ill-fated analysis. By this time however, much work may have been wasted as the original error has been incorporated in later design.

In our environment, the management constraint would appear as a specification of a high level module. It would be verified based on requirements on other modules. These requirements must be matched by specifications on those parts which, in turn, must themselves be verified on the basis of their part requirements. The part module must achieve those requirements through its own specifications. Now, when a low level change occurs that violates the management constraint, it will appear as a tracking action on the specifications of that low level part module. The error is detected immediately.

The designer can either choose to change his/her design accordingly or to negotiate for some changes with users. This is done by changing the offending specification thus causing notification of the involved user module. The conflict could be resolved at this level, or perhaps more changes would be made causing the involvement of the highest level module. In any case, the current state of the design is always evident and delay and wasted effort are effectively eliminated.

Conflicts can appear as a result of the change of any element of the design. During interactive design procedures, engineers are used to being given feedback regarding the correctness of the operations they perform. This is a simple form of verification tracking and notification where the verifications are of design. By incorporating verification tracking in the underlying data base mechanisms, new dimensions are added to this process. By saying that verifications are associated with design items, we have in effect, explicitly separated verification from input process. In this example, feedback is still provided at design time, but now, those verifications are also associated with the resulting design data. Now, when changes are made to components of the design operation, those original verifications will be automatically invalidated. The designer is notified that reverification of the design operation needs to be performed. Instead of receiving design feedback only at design time, design validity is maintained through all aspects of design generation and change. Every kind of design operation can be involved in automatic conflict detection and resolution thus maintaining overall project integrity.

5. ONGOING RESEARCH

We have implemented the data base technique described above as well as the coordination mechanisms verification tracking and notification. The ACE (Automated Computer Engineering) system has been implemented within the ASK (A Simple Knowledgable) environment of Drs. Frederick and Bozena Thompson at the California Institute of Technology. A small example of ACE activity is provided at the conclusion of this paper. Other areas affected by the project structure approach are also being investigated.

One interesting result of this approach is the suggestion that verification processes which usually operate on large projects should be subdivided along lines parallel to the project structure. Exploitation of hierarchy yields savings through redundancy elimination. Subdivision of a process into partial segments performed at times associated with the design process, results in the use of fractional data sets. That is, only a small portion of the data need be analyzed at any one time. Composition portions include the results of element verification. Less data at one time implies less disk usage and therefore better total execution time. Time savings may also accrue as a result of hiding partial verification behind the interactive design process.

Better automatic documentation systems are also sure to result from the project structure approach. A hierarchical decomposition of design is available. Beyond this, we have also come up with a new use for the concept of transformation. Automatic transformations have typically only been written to translate between representations.

Another kind of automatic transformation is one which extracts semantic information from the engineering data. A designer can choose a set of transformations to apply to his or her data in order to generate a particular form of feedback desired. For example, one might define a transformation called "boundary", another called "red" and another called "bearings" each of which extracts a particular set of information from the data base. By concatenating these transformations a designer can receive all kinds of feedback. For example, by joining all three in the order listed above one would get an answer to the question, "What are the boundaries of all red bearings?" This kind of transformation imposes a semantic interpretation upon an engineering data base. The interpretation of "red" or any other transformation is encapsulated in the program which is that transformation. By using the ASK natural language facilities (1) and implementing transformations as ASK operators, one can actually use English sentences to concatenate transformations as in the above example.

Verification can also make use of concatenated transformations. "Display the first mode of building X minus all rear elevators when struck by an earthquake of magnitude 5" makes use of the transformations "first mode", "minus", "elevators", "rear", "earthquake", and "magnitude 5". Another combination of the same transformations would yield entirely different results. This should indicate the range of possibilities of this approach to transformation. By incorporating this into the project structure environment even more powerful semantically guided operations should be possible.

Finally, the concept of parametric module can be extended to include automatic design systems as they can interact with designers and the project structure. A whole new vision of automatic design is revealed. In it, the instantiation process consists of derivation processes and decision processes. Derivation is the production of design from specification parameters and the respecification of modules based on changes to requirements. It involves producing data directly as well as producing changes to existing data based on user decisions. These changes in direction can be provided by the user after each level of derivation. Thus, automatic design becomes an interactive process in which design may occur as a top-down, bottom-up, or inside-out process.

6. CONCLUSIONS

The project structure existing within every engineering organization implicitly provides mechanisms of design team independence. Primarily, this is accomplished through the establishment of requirements on parts which represent the assumptions being made about the make-up of those modules. Exploitation of this structure in the organization of engineering data and systems yields a broad range of potential benefits. These potentials should be especially useful in engineering environments of considerable complexity where status keeping, integration, inter-communication and conflict resolution are serious and well established problems. We have at this time constructed preliminary prototypes for integrated circuit design, mechanical design, and software engineering environments. We hope to assist in the construction of an industrial environment in the near future.

7. SHORT EXAMPLE

>new copter fact (starting a design dialogue)
>Part of the design of what module? block
>Element of specification? socket
>Side? right
>Position? 25,10,15
>Array dimensions? 8,2

The following new specification has been added:

1) Type is SOCKET.
   Direction is RIGHT.
   Position is 25,10,15.
   Array size is 8,2.

The verification of type "reqcon" of requirement 2 of module "landing gear" has been reverified and SUCCEEDED.

The verification of type "compcon" of specification 1 of module "block" has been invalidated.

The verification of type "reqcon" of requirement 1 of module "rear body" has been reverified and FAILED.

(excerpt from module design utility dialogue)
>Action? VS (Verify a specification)
>Number of specification? 1

Can't perform "compcom" verification until "nums" is valid.

***** Verification type "compcom" FAILED.
>Action? AVS (Add a verification)
>Number of specification? 1
>Add what verification? nums
>Add what verification?
Changes saved.
>Action? VS
>Number of specification? 1
*****Verification type "nums SUCCEEDED.
*****Verification type "compcom" SUCCEEDED.

(Using ASK English)

>What is the type and designer of each part of MM?

part	type	designer
inport	port	Segal
left wing	wing	Smith
outport	port	Segal
frame	casing	Segal

>What is the optimistic time, estimated time and pessimistic time of each task of frame?

task	optimistic	estimated	pessimistic
framplan	14	21	28
desorg	10	12	13
finjob	25	36	45
desdins	5	7	8

8. REFERENCES

(1) Dr. F.B. Thompson and Dr. B.H. Thompson, "Introducing ASK, A Simple Knowledgeable System," Caltech Computer Science technical memorandum 5054, 1982.

(2) Richard Segal, "Design Automation and Engineering Organization Structures," ACM SIGDA Newsletter, Sept. 1983.

(3) William Beeby, "The Future of Integrated CAD/CAM Systems: The Boeing Perspective, IEEE Computer Graphics and Applications, Vol 2, Number 2, January 1982.

9. BIOGRAPHY

Richard Segal was born in New York City in 1958. He moved to Albuquerque, New Mexico in 1971 and to California in 1975. There he received a B.S. in Information and Computer Science from the California Institute of Technology in Pasadena in 1981. Since then, under the direction of Dr. Frederick Thompson, Richard has been completing a Ph.D. program in Caltech's computer science department. The expected thesis title is "Computer Integration in Engineering Design and Management." Richard's background includes fourteen years programming experience, integrated circuit design, design tools, and artificial intelligence. He is a member of the ACM, IEEE, SCS, and AAAI.

19

STRATEGIC PROJECT METHODOLOGY FOR A COMPUTER-AIDED ENGINEERING DESIGN SYSTEM

Larry D. Sieck
President
Pacific CAD, Incorporated
El Cajon, California

Abstract

Many federal agencies and military establishments are investigating computer-aided engineering design (CAED) systems. Some agencies and installations are trying to make CAED systems operational and productive. If their ventures are to be successful, a systematic project methodology needs to be utilized.

This paper will present a strategic project methodology currently being used on an Army telecommunications project at Fort Huachuca, Arizona.

The paper will discuss each of the methodology phases and their related key action steps. The paper will show how each phase is designed and built to be a refinement of all previous phases.

1. INTRODUCTION

In 1977 the Army Communications Command adopted a new program for upgrading the telecommunications system on each base which utilized digital switches in a distributed network concept. Initially an attempt was made to implement the new program using existing manual techniques. However, because of the massive amounts of data that needed to be collected and analyzed, automated techniques were explored. In 1980, a group in the Voice Branch of the Communications Engineering Directorate began experimenting with a graphics system to determine its usefulness in the telecommunications engineering effort. By 1983, a prototype system had been developed which proved that the computer graphics approach was the most timely and cost effective method for engineering telecommunications systems.

Before implementing a full production system, it was decided to conduct a review of the total telecommunications engineering requirements and the availability of "off-the-shelf" systems to meet these requirements. This paper presents the project methodology which is being used at Fort Huachuca for the telecommunications engineering study. The methodology contains four phases with related key action steps. The methodology outlined in this paper can be used by any discipline contemplating the acquisition of a CAED system.

2. INVESTIGATION PHASE

The investigation phase at Fort Huachuca has three key action steps: (1) Scope, (2) Data Collection and (3) Cost/Benefit Analysis.

2.1 SCOPE

The project's scope need to be clearly defined and reviewed with management. In essence, the project's scope defined the boundary or "limits" of the study. At Fort Huachuca, the scope substantially involved the automated preparation of a telecommunications engineering installation package (EIP) for both inside and outside plant.

2.2 DATA COLLECTION

In general, this step involved answering the questions "who, what, where, when and how" in regards to the EIP package. In order to answer the above "five questions", a "dossier" was prepared on each drawing and record identified during the study. The dossier format insured that all data was collected and evaluated in a uniform and consistent manner. The dossiers were also valuable in determining input versus output documents and which documents were to be updated.

This step also included obtaining information from users, the availability of existing software and the capabilities of current CAED systems. The "data collection" in this case involved visiting users who had made significant progress with CAED for telecommunication systems. In addition, information was obtained from CAED vendors on their telecommunication CAED capabilities.

2.3 COST/BENEFIT ANALYSIS

Initially, this step involved estimating the cost to prepare an EIP package using manual techniques. The manual costs included data collection, engineering and EIP preparation. The manual costs were then projected over a five-year period. This projection became the baseline to which the CAED alternative was compared.

The cost for the CAED alternative was based on the results obtained from the in-house prototype system. The CAED alternative was also projected for a five-year period. A comparison of the baseline and CAED alternative presents the magnitude of savings to be realized with the CAED system. From the cost comparison it was also easy to obtain the cost/benefit ratio and break-even point.

3. PROJECT PLAN

The next phase of the study will address the project plan. The project plan will define the goals/objectives to be achieved. Each task will contain a narrative description and the time required for completion. While the tasks could be displayed on a CPM or Pert, it is planned that a precedence chart will be used. The project plan will provide the overall direction for the study and include a priority assignment for all EIP applications.

The Project Plan phase has four key action steps: (1) Project Definition, (2) RFP Preparation, (3) Proposal Evaluation, and (4) Contract Negotiation.

3.1 PROJECT DEFINITION

Each application to be implemented will be defined in detail as to its content, uses and users, sources, relation with other applications and attribute relationships. Sample graphic products and/or reports will be prepared showing format, symbology, text sizes, fonts, as applicable.

3.2 RFP PREPARATION

The request for proposal (RFP) must concisely spell out what hardware and software capabilities are required to generate the EIP documents. The technical portion of the RFP will be based on the data obtained in the project definition phase. The evaluation documents will be prepared before the RFP is sent out. A comparison will then be made to verify that all of the evaluation requirements are contained in the RFP itself. A preproposal conference may be held since there are some unique applications and/or special requirements which may necessitate developing "hybrid" systems.

3.3 PROPOSAL EVALUATION

Ideally, the proposals will be evaluated using a "weighted ranking" as opposed to the "pass/fail" process. The advantage of the weighted rankings is that it provides a method for comparing the hardware/software capabilities of each proposer. A key part of the evaluation will consist of preparing a list of questions pertaining to each response to review with the finalists. We plan to visit each of the finalists and address the areas of concern at length. The risks associated with each of the finalists will also be documented and evaluated as to the impact on the project should the "risk" occur, i.e., the vendor goes out of business.

3.4 CONTRACT NEGOTIATION

This is the last opportunity to clear up any questionable areas. Each item of concern will be gone over again and agreed upon in the RFP, the vendor's proposal and a document stating negotiated items will be attachments to the contract.

4. SYSTEM IMPLEMENTATION

There are three key action steps associated with system implementation.

4.1 SITE PREPARATION

There are numerous substeps associated with site preparation. Depending upon the system obtained, a special air conditioned room may be needed for the central processing unit and/or work stations. Power requirement, BTU generation and space considerations need to be factored into the site plans. Fire protection systems and procedures will also be included in the site preparation plans. The route the equipment will travel in the building will be reviewed with the vendor customer service engineer and arrangements made for any special moving equipment.

4.2 SYSTEM ACCEPTANCE VALIDATION

System acceptance tests will be conducted at the vendor factory and again at the Fort Huachuca. Both factory and site acceptance tests will be conducted to verify that the hardware and software are performing according to the technical specifications in the contract. Remedial procedures will be stated in the contract, defining the action the vendor must take to make the system acceptable and in what time frame.

5. APPLICATION IMPLEMENTATION/ REVIEW

Applications will be implemented using the priority established in the project plan. Each application, depending on its complexity/difficulty, will have its own implementation timetable.

5.1 INITIAL/ONGOING PROCEDURES

Each application will need to be "debugged" before it can be fully implemented. User training programs will have to be established, procedures manuals written and orientation provided to appropriate areas for each application.

5.2 SERVICE BUREAU INVOLVEMENT

Numerous manual drawings and/or records will have to be converted before the EIP program can be executed. Using a quality service bureau may be the fastest and most cost effective method for converting the regional data. Similar procedures to those described under RFP Preparation, Proposal evaluation and Contract Negotiation will need to be followed in selecting a service bureau. Particular attention will be paid to defining product acceptance and when the service bureau can except payment for completed work.

5.3 PROJECT REVIEW

At specified intervals, the project will be reviewed to determine whether the original goals/objectives have been attained. Conducting timely periodic reviews of the project will allow time to take corrective action, thereby increasing the probability of project success.

6. CONCLUSION

Justifying, evaluating and implementing a CAED system is not necessarily difficult, but there are many tasks which must be completed in a logical sequence. This paper attempted to show a strategic project methodology which will result in a cost effective project which achieves its stated objective in a timely manner.

7. BIOGRAPHY

Mr. Sieck is the President of Pacific CAD, Inc. Since 1980 he has been a CAD consultant assisting the federal, public and private sectors.

Prior to 1980 he was responsible for implementing a CAD system used for utility drawings, and developing another intelligent CAD system for engineering and mapping applications.

Mr. Sieck has a Bachelor of Business Administration degree from National University where he graduated Summa Cum Laude.

20

DEMAND
THE FIRST OPERATIONAL OPTICAL DISK BASED DEMAND PRINTING SYSTEM

Lee F. Wilbraham
Manager, Advanced Systems Concepts and Planning Department
Xerox Special Information Systems
250 North Halstead Street
P. O. Box 7018
Pasadena, California 91109 U.S.A.

Abstract

DEMAND is a computer-driven optical disk system. It stores images, retrieves and prints on demand. The Library of Congress (LOC) Cataloging Distribution Service (CDS) uses DEMAND to capture, store and deliver their 5.5 million non-MARC active library cards. This paper describes the DEMAND system, its major subsystems and recently developed equipment. In addition, the paper discusses technology progress since the DEMAND system was dedicated in August 1982.

1. INTRODUCTION

The purpose of this paper is to describe the Library of Congress Cataloging Distribution Services DEMAND (Digitized Electronics MARC and Non-MARC Display) system technology. The paper describes the DEMAND system functional overview, discusses each of the major subsystems and components, near-term technology developments and their impact on future document delivery systems like DEMAND.

The DEMAND system is the second phase of the Library of Congress two phase program to automate the production of library cards. The first system, CARDS (Card Automated Reproduction on Demand System), was implemented in 1978. Produced by Xerox Special Information Systems (XSIS), CARDS prints in an automative manner post-1968 Roman Language machine readable (MARC) catalog cards.

With the dedication of the DEMAND system in 1982, LOC has the capability to demand print both machine readable and non-machine readable (Non-MARC) cards. The two systems significantly:

- o Reduce labor required
- o Reduce response time for orders
- o Reduce space required
- o Provide for future growth in services
- o Increase file integrity—Protecting against wear and loss

2. DEMAND SYSTEM FUNCTIONAL DESCRIPTION

The Library Catalog Distribution Service formerly stored seven million 3 by 5-inch catalog cards in their original card stock form. When a public library or other organization requested certain catalog cards, the original card was located in the database, copies were run and distributed, and the original card was returned to the file. See Figure 1 for a flow chart of this process. This large file is characterized by the following attributes:

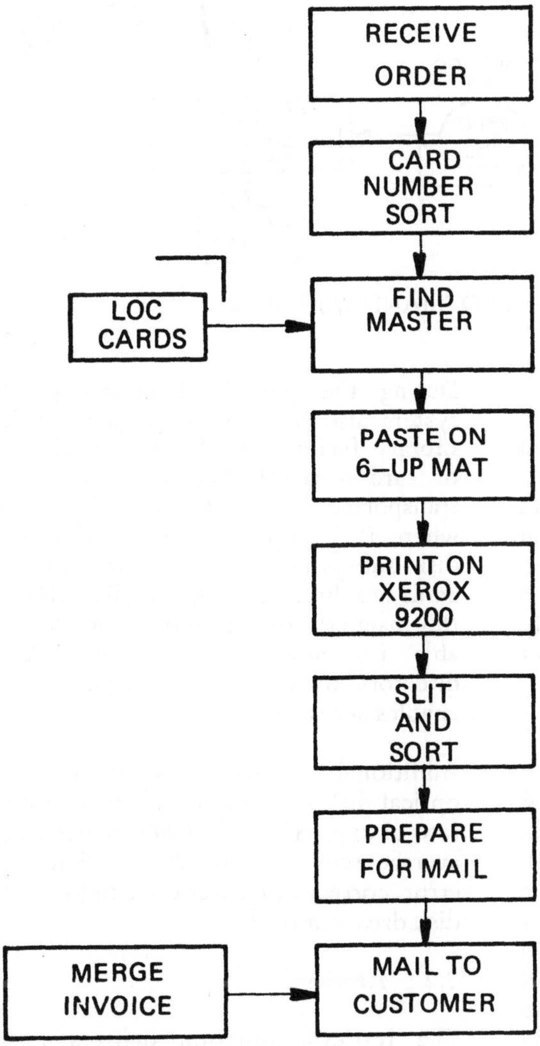

Figure 1. Manual Work Flow Diagram

- o A major portion is non-Roman format.
- o Fonts vary extensively over the 70 year collection period.
- o The manual processes are labor-intensive.
- o Due to physical handling, file integrity is compromised.

2.1 DEMAND SYSTEM

The DEMAND System provides the means for reducing manual file handling, floor space requirements, order processing time, and the number of personnel required to provide the service. The DEMAND system concept is to store high-resolution digital images of catalog cards on mass storage devices and to retrieve and print customer-requested cards utilizing a high-speed laser xerographic printer. This is accomplished by employing an electronic storage and retrieval system for the following functions:

- o Capture
- o Verification
- o Mastering
- o Retrieval

A work flow diagram is shown in Figure 2.

2.2.1 Capture

The capture function provides the facility for scanning the original cards at 480 bits-per-inch resolution. Each card is scanned and then displayed on a high-resolution video terminal. An operator views the image, accepts it qualitatively or rejects it by initiating a re-scan, and provides a unique Library of Congress key through a data entry keyboard response. The resultant image is compressed and transferred to temporary magnetic storage. An approximate compression ration of 20:1 is achieved through a combination of run-length encoding and two-line prediction algorithms.

2.2.2 Verification

The verification function routes a successfully scanned card image to another video terminal, where it is decompressed and displayed. As prompted by the system, an operator enters the card key, and the system programmatically compares this with the previously supplied key. Retries are permitted when a match does not exist.

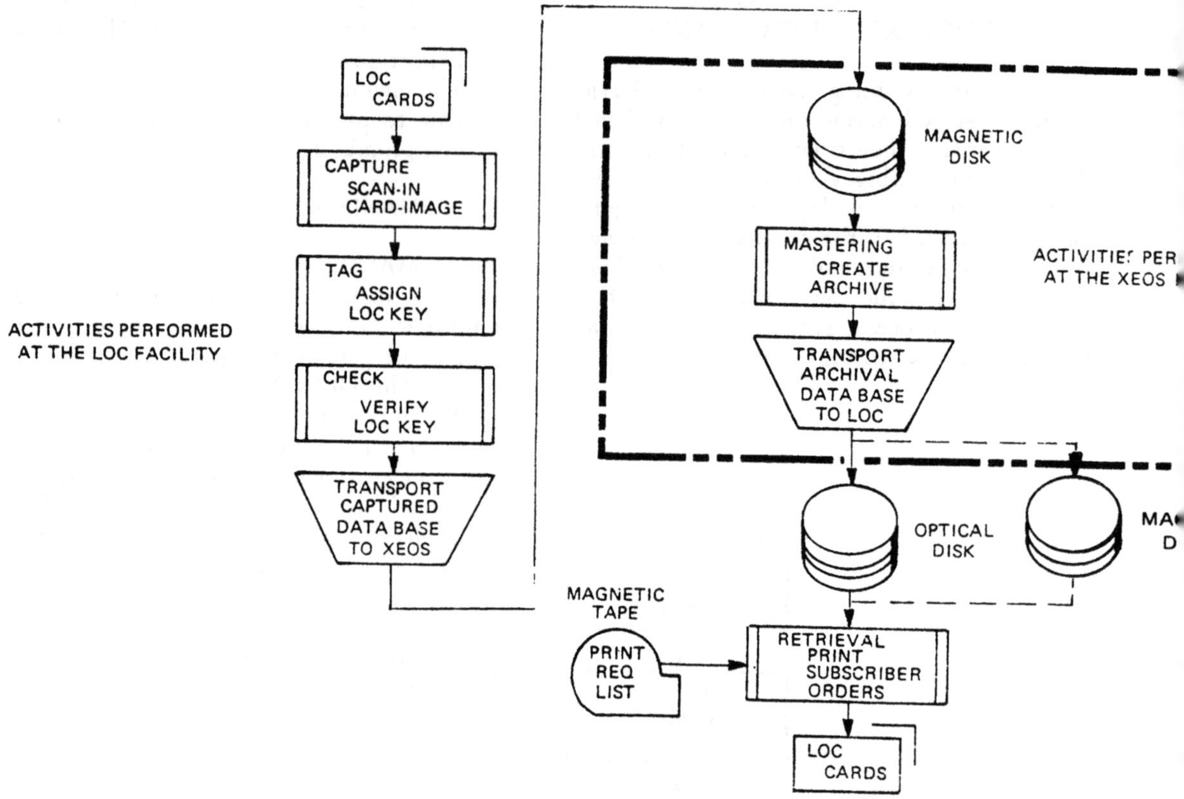

Figure 2. Overview of DEMAND Work Flow

2.2.3 Supervisory

Conflicts remaining between the captured and verified keys are resolved through the supervisory function. Having access to both data entry responses, the supervisor acts an an arbiter to determine proper card identification. Conflicts can be deferred and handled in a batch processing mode or resolved interactively by allowing a lead verification operator to act as the supervisor.

2.2.4 Mastering

A file system and associated index are created for the verified cards on magnetic disk. Duplicate files are copied to magnetic disk pack and transported in batches from the Library facility to the XSIS facility for mastering. Approximately 18 (300 MByte) magnetic disk packs are accumulated, and their contents copied to one 4.5 GByte optical disk. An index describing the location of the cards in the archival file is also written to optical disk.

During the period of mastering, the file system also remains on magnetic disk at the Library facility to fully provide the function of card retrieval. The mastered cartridge is transported from XSIS to the Library facility where it is mounted on a read-only optical disk subsystem. At that time, requests for cards are directed to optical disk, thus making the magnetic disk resource completely available for new scanning tasks. A master directory maintained on magnetic disk pack enables access to all cards.

Additional characteristics of the read-only optical disk subsystem include a data transfer rate of approximately 1 MByte per second, an average seek time of 250 milliseconds, and error correction/detection employed by the disk drive controller.

2.2.5 Retrieval

The retrieval function delivers images of requested cards from the database to a video terminal for operator perusal or builds a magnetic disk file of cards to be printed by a modified 9700 electronic printer. When a requested card image resides on a disk in the off-line library, the system prompts the loading of that disk to the on-line state. The printing subsystem formats and prints both graphics (card image) and text (subscriber information) for six cards per page at a resolution of 480 bits-per-inch.

3. SYSTEM IMPLEMENTATION OVERVIEW

A block diagram of the three DEMAND subsystems (Input, Retrieval, and Mastering) is shown in Figure 3. The subsystems are co-located. The Mastering Station is located in Pasadena, California, at XSIS. The Input and Retrieval Subsystems are located at the LOC but in different areas.

3.1 INPUT SUBSYSTEM

The DEMAND Input Subsystem (Capture/Tag/Check) generates compressed bit-stream representations (images of library cards) and produces:

- o Disk packs on which complete and correct index information is associated with each card image
- o Statistical information in the form of reports and tapes

The Input Subsystem allows any combination of functions (Capture, Tag, Check, Supervisor) to reside on any number of disks (from one to four) and provides for run-time reconfiguration of these functions in relation to the on-line disks. The Input Subsystem also provides control options for system start/restart, selective reconfiguration, and halt.

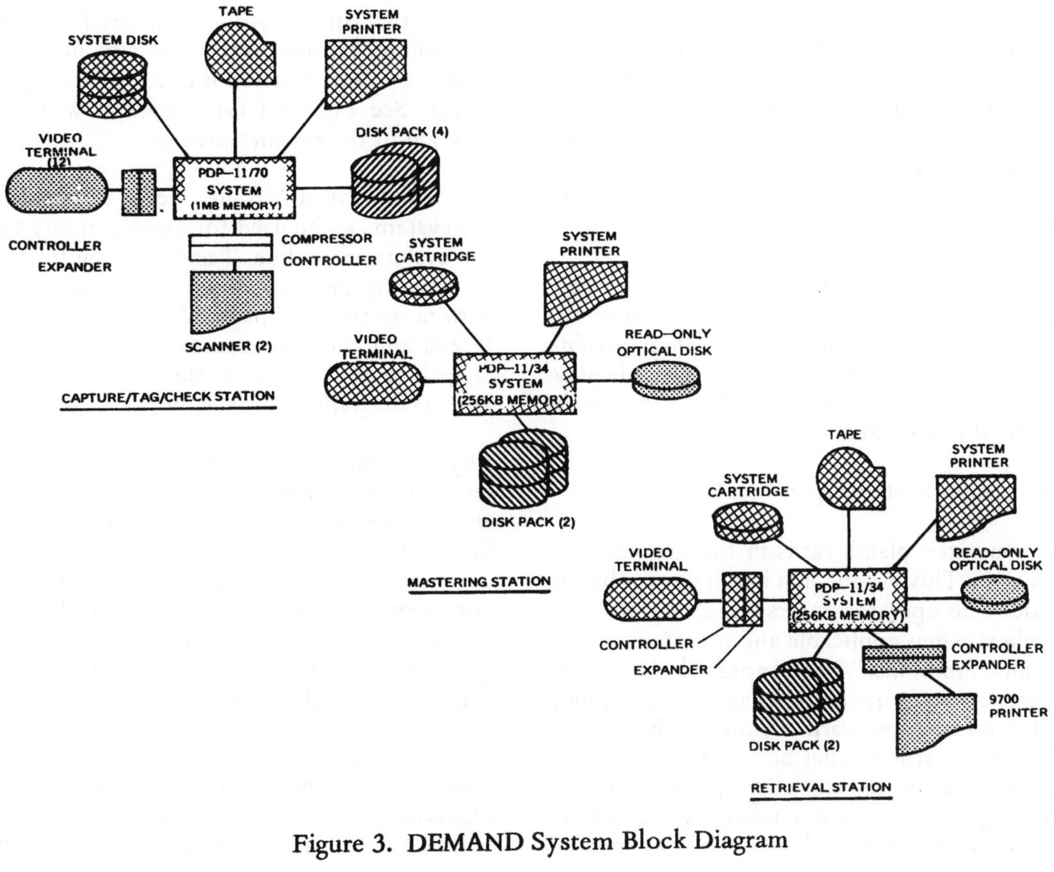

Figure 3. DEMAND System Block Diagram

The system consists of the following hardware:

- o DEC PDP 11/70 with 1 MByte memory
- o DECwriter, which functions as computer operator's console
- o One 67 MByte system disk
- o Four 300 MByte magnetic disks for storage of up to 11,000 plus images per pack
- o Two scanners for the inputting of images in compressed format
- o Two sets of six 1024 by 1024 terminals each for use as either Capture, Tag, Check, or Supervisor Stations with keyboards
- o Magnetic tape drive for creation of tape files

Scanned images are compressed by hardware in the scanner controller. The compressed images average about 20:1 compression. The system processes the images in compressed format. The images are only expanded when they are displayed or printed. The display and printer both have hardware decompressors. The scanner is described in detail in Section 4.

The DEMAND System compression algorithm is a modification of the British Post Office algorithm. Both the expander and compressor are very high speed units because of the 480 pixel per inch requirement. The expander operates above 20,000,000 bits per second. Both were developed on the DEMAND program.

The card image display was also developed under the DEMAND program. It contains a controller, data decompressor and high resolution CRT with a keyboard. The display resolution is 1024 by 1024. The monitor is mounted in a vertical orientation.

3.1.1 Input System Operation Overflow

The operator places cards in the scanner one at a time. This is done in a batch or non-batch mode. The operator enters continuation card numbers when applicable and initiates scan or rescan functions. The captured image is placed on a tag queue indexed by date, time, and confirmation card number. The tagger adds LOC card number and additional index information or rejects poor quality images. The tagged images are verified by a checker. Problems are resolved by the Supervisor. When completed, the images are placed in a queue for distribution to the computer operator for the final processing. The final processing includes separating the bad images, sorting the good image index, creation of a backup copy of the good image disk, and creation of the good image disk index tape. This process results in:

- o The good image index tape is sent to the LOC Data Base Directory
- o The backup good image disk is available for the retrieval process
- o The good image disk is used for the mastering process

3.1.1.1 Capture, Tag, Check. There are twelve terminals. Any two of the twelve terminals may be used for capturing images at any time the system is on-line. At each station an operator logs on, enters a number of LOC cards into the DEMAND System, and then logs off. The station is capable of running in either batch or non-batch mode. For each LOC card entered, the capture station associates some additional information. In batch mode, that information is entered once for the whole capture session and in non-batch mode, that information is entered later by the tagger. See Figure 4 for a work flow process diagram of the capture process.

Any combination of the twelve terminals on the system can be used for tagging at any time the system in on-line. Tagging is the function of entering additional data to be associated with a captured image. After an image is tagged and checked successfully, it cannot be tagged again. Figure 5 shows a work flow process diagram of the tagging process.

Any combination of the system's twelve terminals may also be used for checking at any time the system is on-line. Checking is the function of verifying that the tagging was done correctly and correcting the data if it is not correct. After an image is checked and tagged successfully, it cannot be checked again. See Figure 6 for a work flow process diagram of the checking process.

3.1.1.2 Supervisor. One of the twelve terminals on the system can be used for supervisory functions any time the system is on-line. The supervisor station can either

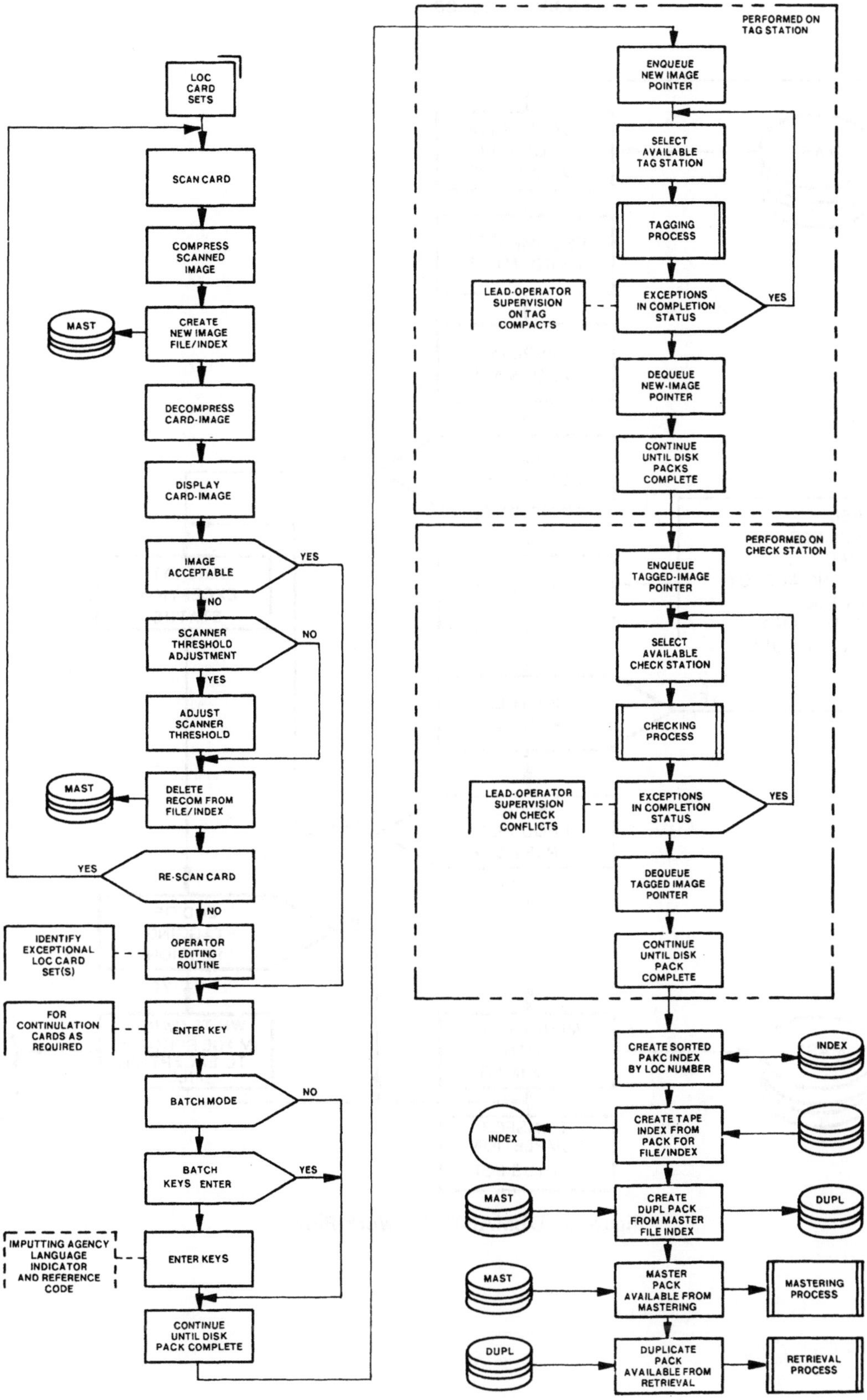

Figure 4. DEMAND Capture Work Flow

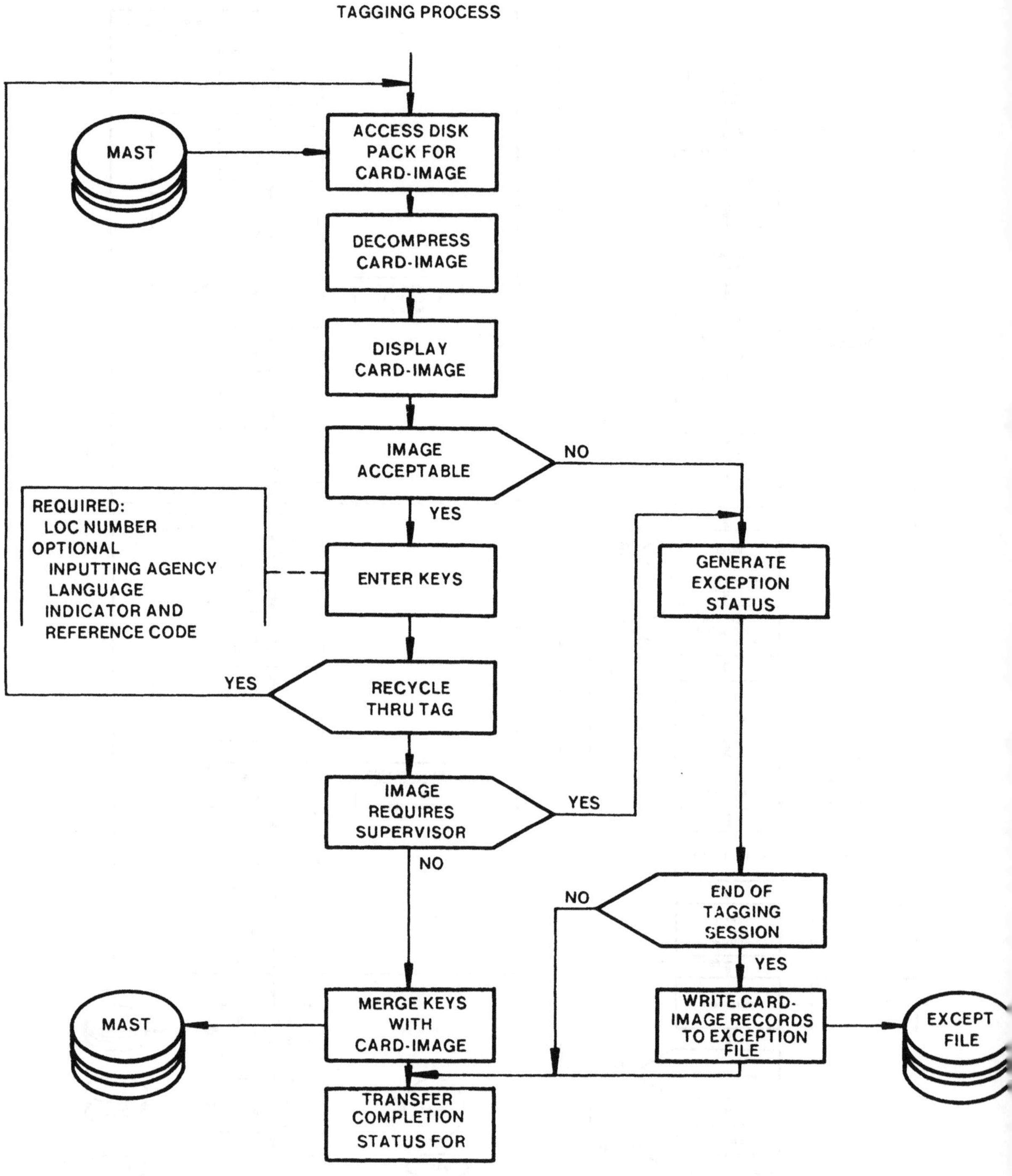

Figure 5. DEMAND Tag Work Flow

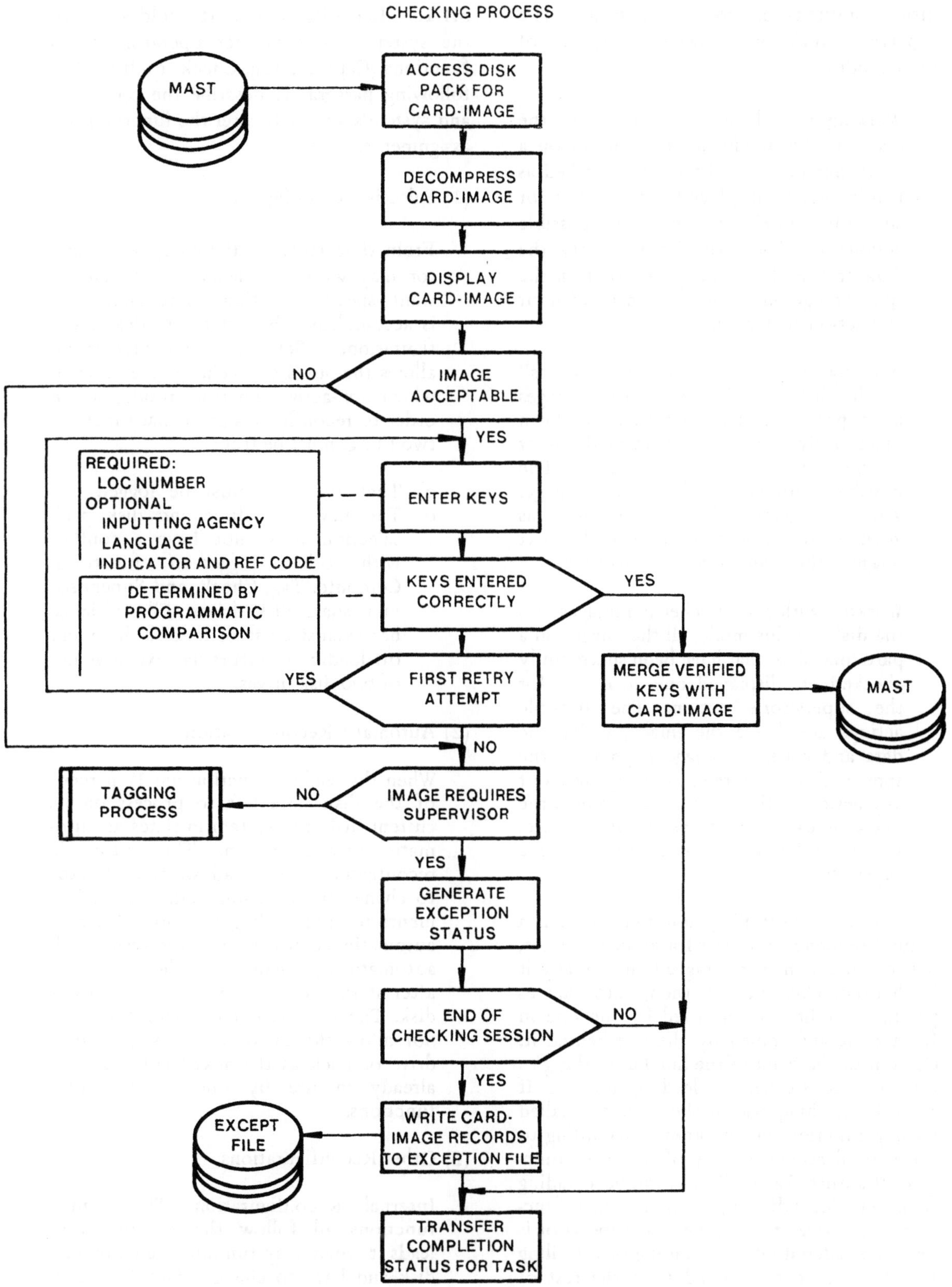

Figure 6. DEMAND Checking Work Flow

display statistics or work with images. The supervisor works with images in any one of three modes:

o Working with all the "bad images" on the disk. In this mode, all the images on a particular disk that have been marked as bad images are displayed one at a time for the supervisor's action. The possible actions are delete the image, verify the data and mark the image as good on the spot, and go on to the next image without any action at this time.

o Working with all the "unverified images" on the disk. In this mode, all the images on a particular disk that have not been successfully checked are displayed one at a time for the supervisor's action. The possible actions are delete the image, verify the data and mark the image as good on the spot, and go on to the next image without any action at this time.

o Working with all the "checked images" on the disk. In this mode, all the images on a particular disk that have been successfully checked are displayed one at a time for the supervisor's action. The possible actions are delete the image, verify the data and mark the image as good on the spot, and go on to the next image without any action at this time. In this mode, the operator can look at every "nth" image, with the value of n at the discretion of the supervisor.

3.1.1.3 Leading/trailing function. There is a natural sequence of events for an image. First, it is captured, then it is tagged, and finally it is checked. Conceptually then, these are as people in a line. Each is led by the one in front of it and trailed by the one behind it. Of course, the front of the line has nothing to lead it so it is called the leading function. If there is anything else in the line, it is called trailing function and is ordered according to the natural sequence. Any of the three functions (Capture, Tag or Check) can be a leading function, depending upon what other functions are active at that time. The supervisor is never considered either a leading or a trailing function, so it is excluded from the rest of this section.

3.1.1.4 Reconfiguration. At "cold start" of the system, the computer operator assigns functions (Capture, Tag, Check) to disks. The following paragraphs describe the operator's and system's ability to alter the function/disk assignments.

(1) Selective Reconfiguration

From time to time, the computer operator may wish to specify which disk he wants specific functions to be using. This is accomplished by using selective reconfiguration. Selective reconfiguration allows the operator to change the disks on which non-active functions reside, i.e., in order to reconfigure a particular function, two things must be true:

o That function must be logged off.
o The new disk drive and disk pack selection must not have a conflict with other logged-on functions. Compute, Tag, Check, and Supervisor may share disk drives. File initialize has exclusive use of its disk drive and off loading of errors has exclusive use of two disk drives.

(2) Automatic Reconfiguration

When the leading function has 10 percent of the disk data left to process on its current disk, the system initiates an automatic reconfiguration. The automatic reconfiguration only allows the operator to change the disk unit and pack assignments for the leading function. After an automatic reconfiguration, the system will automatically change to the new disk after it runs out of space on its current disk. The automatic reconfiguration does *not* allow the operator to assign a disk drive or pack as the next leading pack is already in use by one of the other functions.

(3) Other Reconfigurations

Internal Reconfiguration. The trailing functions all follow the function that leads it when they run out of data on one disk and have to change disks. The only time they would have no place to go is when they "catch up" to their leader. These reconfigurations (called internal reconfigurations) can happen any time there are trailing functions active.

Supervisor Reconfiguration. When the supervisor wishes to process images on a disk, the supervisor does a reconfiguration from the terminal by specifying a disk drive and ID.

File Initialize Reconfiguration. When the computer operator wishes to initialize an image disk, the operator does a File Initialize Reconfiguration.

Offload Errors Reconfiguration. When the computer operator wishes to offload errors from one disk pack, the operator does an Offload Errors Reconfiguration.

The system will do only one reconfiguration at a time.

3.1.1.5 Unverified Images. Images that were not tagged or checked as correct even with the retries as specified in subsection 3.1.1.7 are called unverified images. These images can only be resolved by a supervisor. They can be moved off one disk pack and onto another disk pack for future resolution. The moving of these images is accomplished by the function of offloading errors.

3.1.1.6 Bad Images. When the capture station indicates that an image is bad, it is never stored to be tagged and checked, but it is possible for a tagger or checker to indicate that an image is bad even though it was considered good by the capture station operator. In that case, the image is considered a bad image. These images can only be resolved by a supervisor. They can be moved off a one-disk pack and onto another pack for future resolution.

3.1.1.7 Tagging/Checking Retry Algorithm. This algorithm defines all retry logic if images are not satisfactorily tagged or checked. This algorithm does not refer to bad images.

- o A tagging station will always get a second chance to tag an image that was not satisfactorily tagged the first time.
- o If one tagging station cannot successfully tag an image, the system will route that image to another tagging station. This means that there are at most four consecutive tries to tag a particular image. If all four fail, then the image is unverified. If any are successful, the image is sent to a checking station.
- o A checking station will only get images that have been successfully tagged. The station will always have a second chance, if necessary, to check the tagging.
- o If an image fails the checking cycle (two tries), then it goes back through the whole tagging cycle for a second time. If it passes the tagging cycle again, then it comes to the check station again. If the image passes the check, then it is verified. If not, then it is an unverified image.

3.1.1.8 Image Sort Key. After an image is successfully checked, it has a valid sort key associated with it. The key is in the form XXYYAASSSSSSCC, where each of the fields is defined below.

- o XX – Century Code. The value in this field depends on the value of the year entered by the Tagger.
- o YY – Year. This is the two-digit year as entered by the Tagger.
- o AA – Alpha Prefix Code. The value in this field depends on the value of the alpha prefix entered by the Tagger. The entered field is five characters, and an internal table is used by the system to convert the entered field to a two-digit numeric code.
- o SSSSSS – Serial Number. This is the six-digit serial number entered by the Tagger and checked by the Checker.
- o CC – Continuation Code. This is the two-digit continuation code entered by the Capture station operator.

3.1.1.9 System Throughput. The system has the following throughput:

- o In the absence of a reconfiguration, a capture station will be capable of processing 240 cards per hour.
- o In the absence of a reconfiguration, a tag station will be capable of processing 180 cards per hour.

- o In the absence of a reconfiguration, a check station will be capable of processing 120 cards per hour.
- o The sorting of disks will be accomplished by the DEC sort package.

3.1.1.10 System Statistics. Certain statistics pertaining to the input subsystem are maintained including items of interest about the terminal, disk, function being performed, and certain audit statistics.

- o Disk Statistics. Statistics pertaining to any disk including disk ID, function being performed, data on images, and data on sectors on the disk are maintained on each disk and output at the supervisor station and/or DECwriter as requested.
- o Terminal Statistics. Statistics pertaining to any terminal including the terminal number, employee ID, function being performed, data on card processing, and any terminal-dependent time data are maintained for each logged-on terminal. They are output at the supervisor station and/or DECwriter as requested by the supervisor. Some statistics are displayed at each terminal at log-off time.
- o Audit Statistics. Statistics recording the activity at work stations are maintained on the system disk. Enough information is recorded to be able to trace the activity on any image in the system. One file will be created on the system disk for each day's activity. The system has the capacity to store more than one week's data without overflowing. If these files overflow, statistical data will be lost. Facilities exist in the system to dump these data to magnetic tape by date and to purge the file by date. It is the responsibility of the computer operator to maintain space in this file.

3.2 OUTPUT SUBSYSTEM

The Output Subsystem prints library cards from images recorded on optical disks which were captured by the Input Subsystem. A work flow process diagram is shown in Figure 7. A print request list tape contains orders for LOC cards prepared on the LOC System/370. This tape serves as input to the retrieval process and is copied to a work disk file. Based on these requests, the card images are retrieved from magnetic and/or optical image disks and written to a work disk file. Both the print request file on the work disk and the index file pointing to the retrieved card images on the work disk are sorted for use in the formatting phase.

Non-MARC cards are printed on 9 by 10 inch card stock with six 3 by 5 inch cards per 9 by 10 inch page. Images are formatted in subscriber sequence on the work disk. Each subscriber order is preceded by a header label card and followed by a trailer label card. The cards are formatted by either of two methods. The 6-up LOC method dictates that there are six stacks (columns) of cards printed simultaneously (see Figure 8). Cards must be constructed on a subscriber basis such that no subscriber order overlaps into another column. The 1-up wraparound method allows subscriber orders to overlap into other columns. A field in the print request list tape header record determines which method is to be used in formatting. After formatting, this work disk containing the cards is then ready for printing and serves as input to the actual printing process.

At the end of formatting the cards on the work disk, an unfilled orders tape is created and a batch summary report is printed. The unfilled orders tape contains the unfilled print requests and a statistical summary record. An unfilled order is the result of a card image not being found on the specified image disk or a result of a lack of space on the work disk for card images. The batch summary report contains a list of all filled and unfilled requests and a totals page.

The retrieval process also provides the capability of creating a skeleton print request tape. This is accomplished by the recalling of card images for viewing on a terminal. As images are recalled, a print request can be made for a card. These requests are only skeleton requests and are put on a work disk. At the end of the recall session, the skeleton print request tape is made from these skeleton requests on the work disk. This tape must return to the LOC System/370 so the requests can be made complete. Thus, this tape becomes a print request list tape and can then be returned to the retrieval subsystem for printing if so desired.

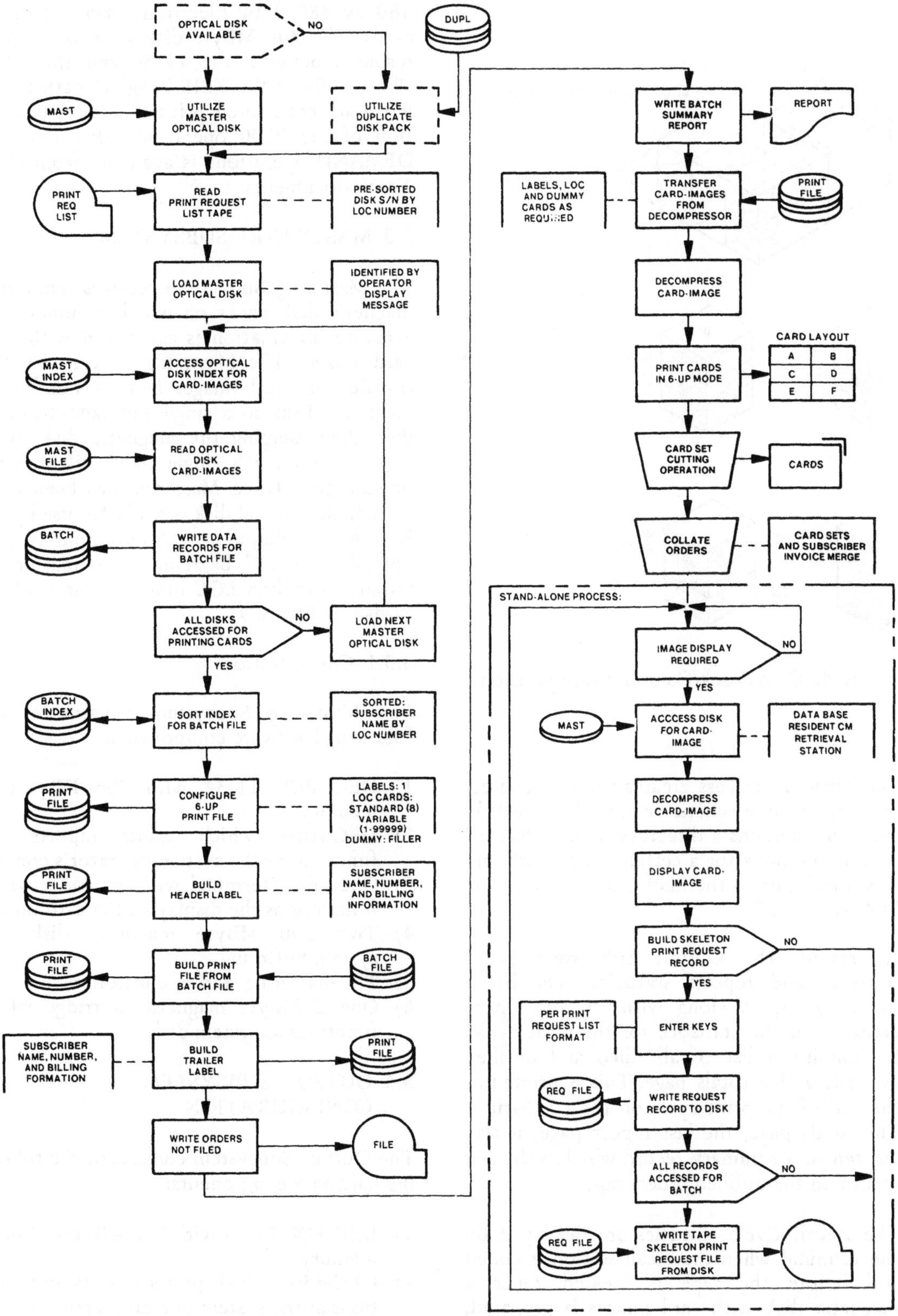

Figure 7. DEMAND Retrieval/Printing Work Flow

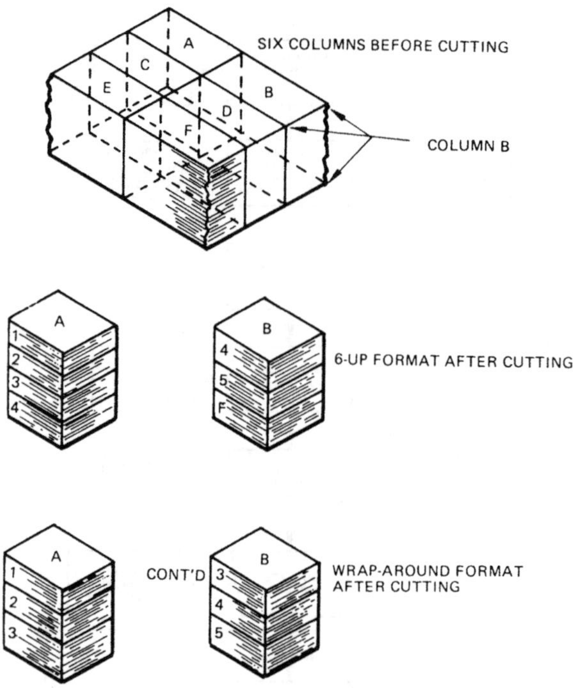

Figure 8. Card Stack/Column Configuration

The retrieval process creates the index directory tape from magnetic or optical image disk. The tape contains a directory of all LOC card catalog numbers for a certain image disk and serves as input to the master directory on the LOC system/370.

As stated, the Output Subsystem keeps statistics and reports statistics. The batch summary report along with its totals page contains all the statistics for the print cycle. It contains a list of all filled and unfilled requests and a totals page. This is printed at the end of the print cycle on the DECwriter. The totals page, the last report page, is also written to a summary record which is the last record on the unfilled orders tape.

The Recall Cycle statistics are displayed on the terminal when the recall session is ended and before the tape is created. Once a magnetic disk with card images is prepared, the printing process may be started. During the printing, the PDP 11/34 works with the 480 by 480 9700 Electronic Printer to print cards. The 300 MByte disk pack is a shared resource between the 11/34 and the 9700. The details of the XSIS imaginal path to the 9700 are considered private data. A description of the 9700 operation other than the DEMAND operations is available from other Xerox documents.

3.3 MASTERING SUBSYSTEM

The Mastering Subsystem receives temporary magnetic disk packs on which complete and accurate information is associated with each card image. The Mastering process involves transferring card images from a number of magnetic disks to a single permanent optical disk then merging the magnetic disk index files to produce a single index file for the optical disk. Once Mastering has been completed, the optical disk is ready for use by the Retrieval subsystem. Approximately 25 optical disks will be required to master the initial 5.5 million LOC non-machine readable cards (see Figure 9).

3.3.1 Configuration

The Mastering Subsystem consists of the following hardware components:

1) DEC PDP 11/34 with 256 KBytes of memory
2) DECwriter which prints reports and functions as the system operator's console
3) One video terminal with expander which functions as the display of the card images
4) Two 300 MByte magnetic disks for system buffering
5) The mastering station controller
6) One 5 MByte magnetic cartridge which functions as system disk

3.4 OUTPUT SUBSYSTEM CONFIGURATION

The Output Subsystem consists of the following hardware components:

1) DEC PDP 11/34 with 256 KBytes of main memory
2) DECwriter which prints reports and functions as the system operator's console
3) Terminal which functions as the recall cycle console

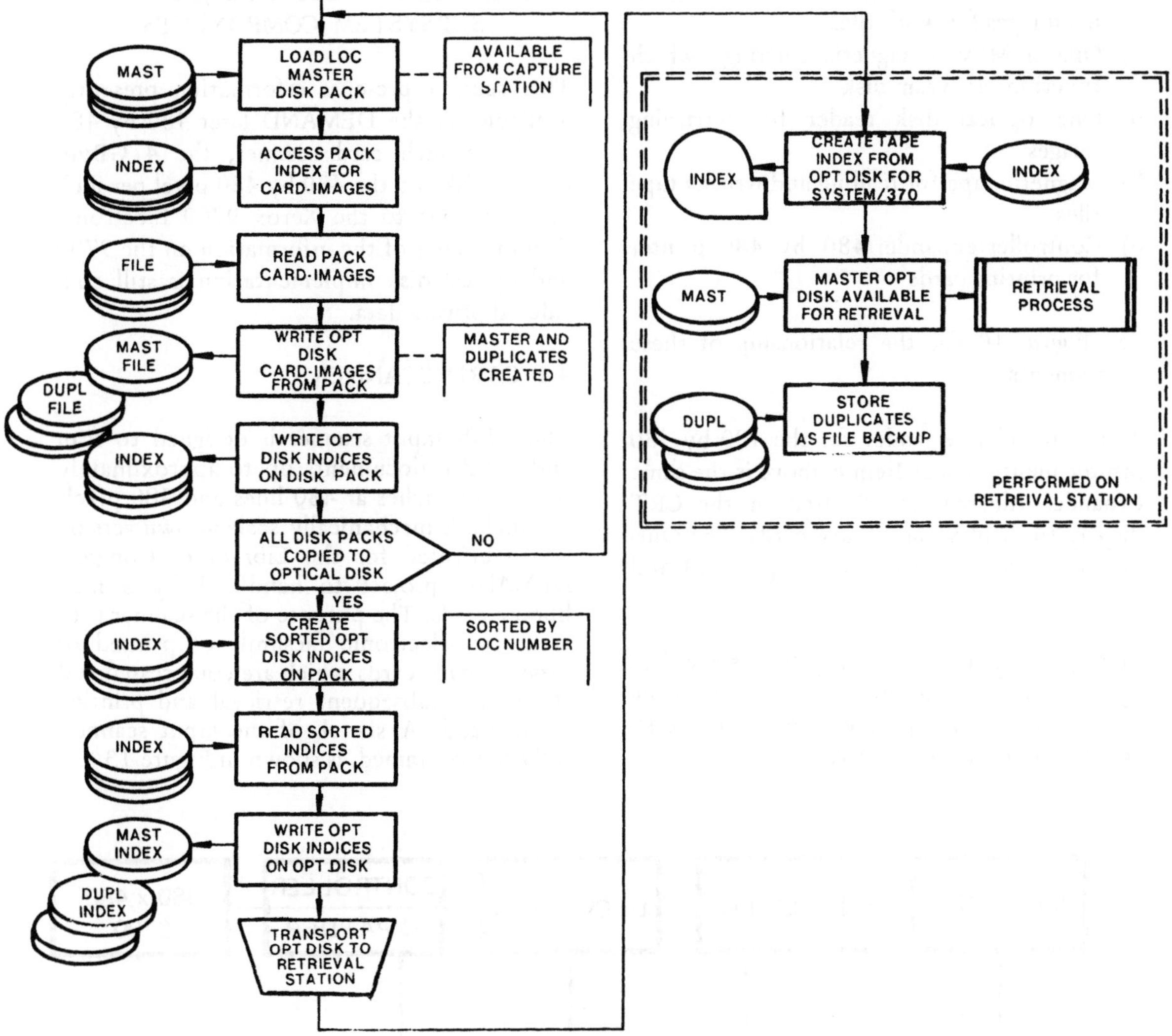

Figure 9. DEMAND Optical Disk Mastering Work Flow

4) Two 300 MByte magnetic disks for retrieving images for work files
5) One 5 MByte magnetic cartridge which functions as sytem disk
6) One optical disk reader for retrieving images
7) Magnetic tape for reading and writing tape files
8) Controller/expander/480 by 480 printer for printing cards

See Figure 10 for the relationship of these components.

The Controller/expander for the 480 by 480 printer mentioned as Item 8 above is the same expander (decompressor) used in the CRT display to display the library cards. The Output Subsystem is pictured in Figures 11 and 12.

The optical disk in this subsystem is the first operational optical disk in an operating environment. The specifications in the disk are presented in Section 4.2.

4. DEMAND SYSTEM MAJOR NEW SUBSYSTEM COMPONENTS

This section presents information presently available on the DEMAND laser 480 by 480 pixel per inch card scanner, the 4 GByte optical disk and the 480 by 480 pixel per inch modifications to the Xerox 9700 Electronic Printer. Much of the information on the 9700 and optical disk implementation is still considered private data.

4.1 CARD SCANNER

The XSIS input scanner is designed to scan and digitize documents up to approximately 8½ by 11 inches at 480 lines and 480 pixels per inch. A mechanically scaled-down version was fabricated for the Library of Congress DEMAND project to handle 3 by 5 inch library cards. The purpose of the scanner is to create an electronic facsimile of printed or typed library cards which are compressed and stored for subsequent retrieval and printing on demand. A sketch of the input scanner, fully self-contained, is shown in Figure 13.

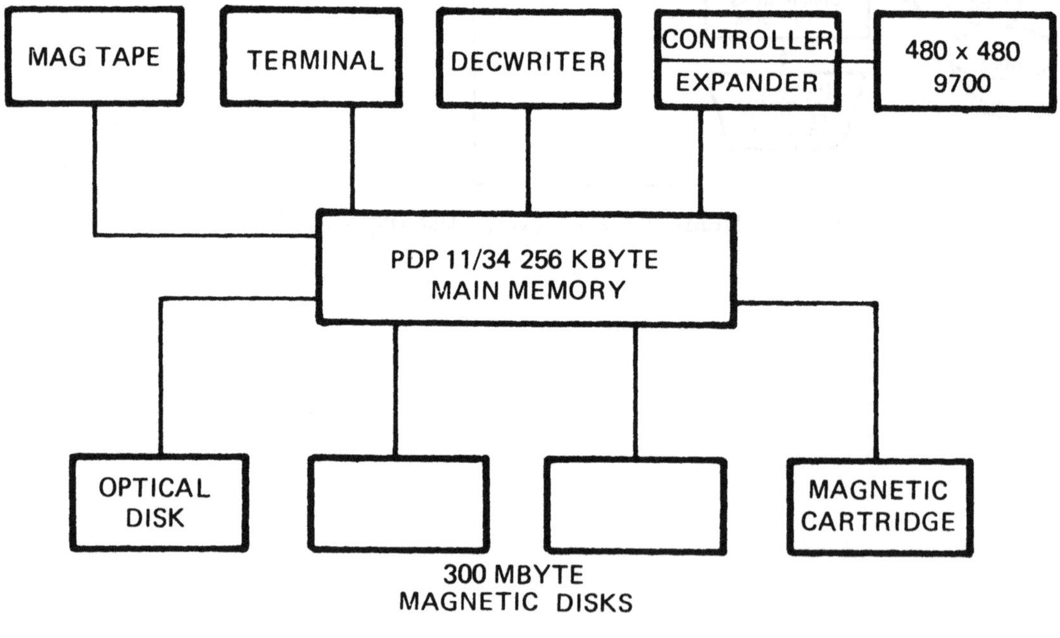

Figure 10. Output Subsystem Configuration

Figure 11. Library of Congress Output Subsystem Computer with Optical Disk

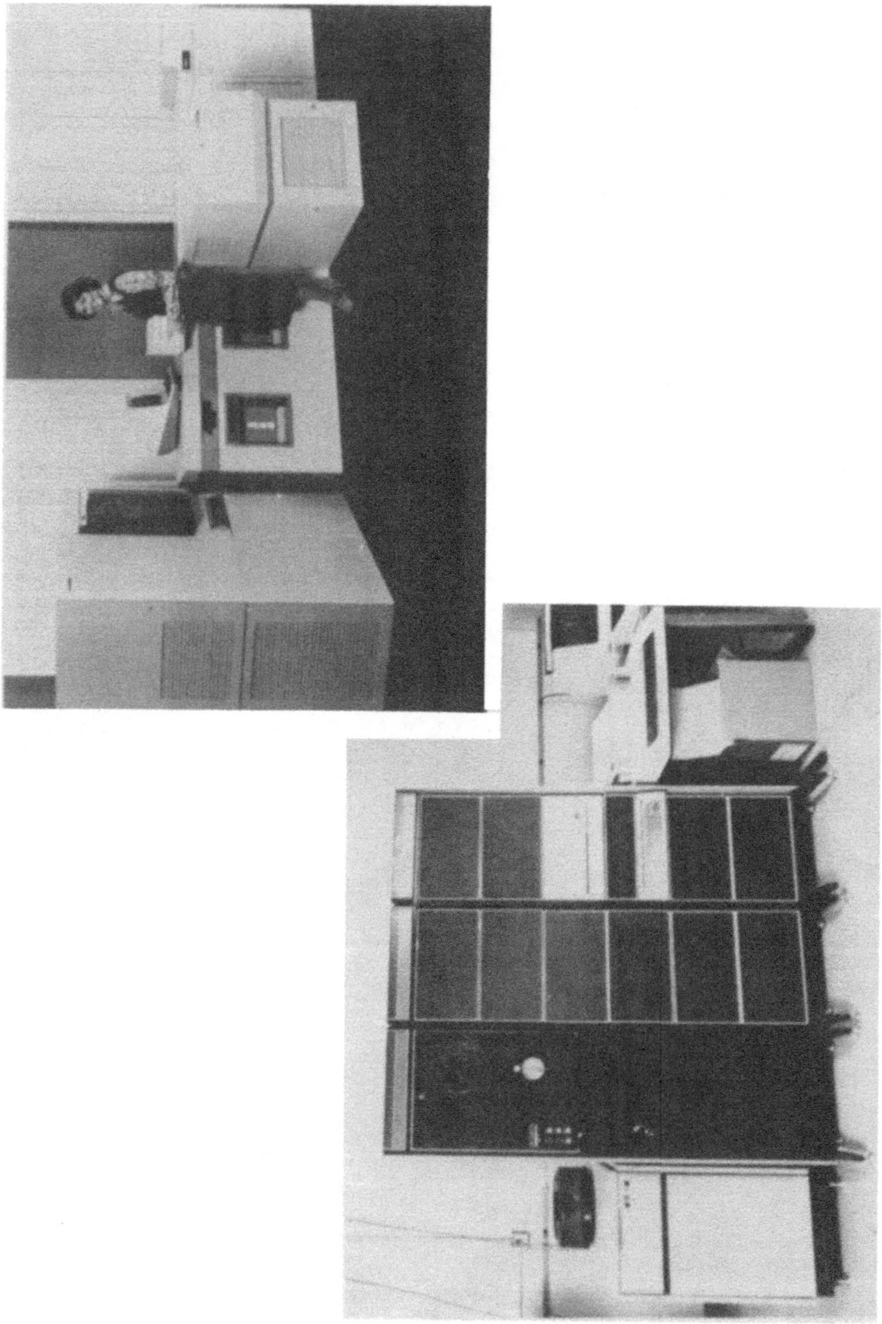

Figure 12. Output Subsystem with Printer

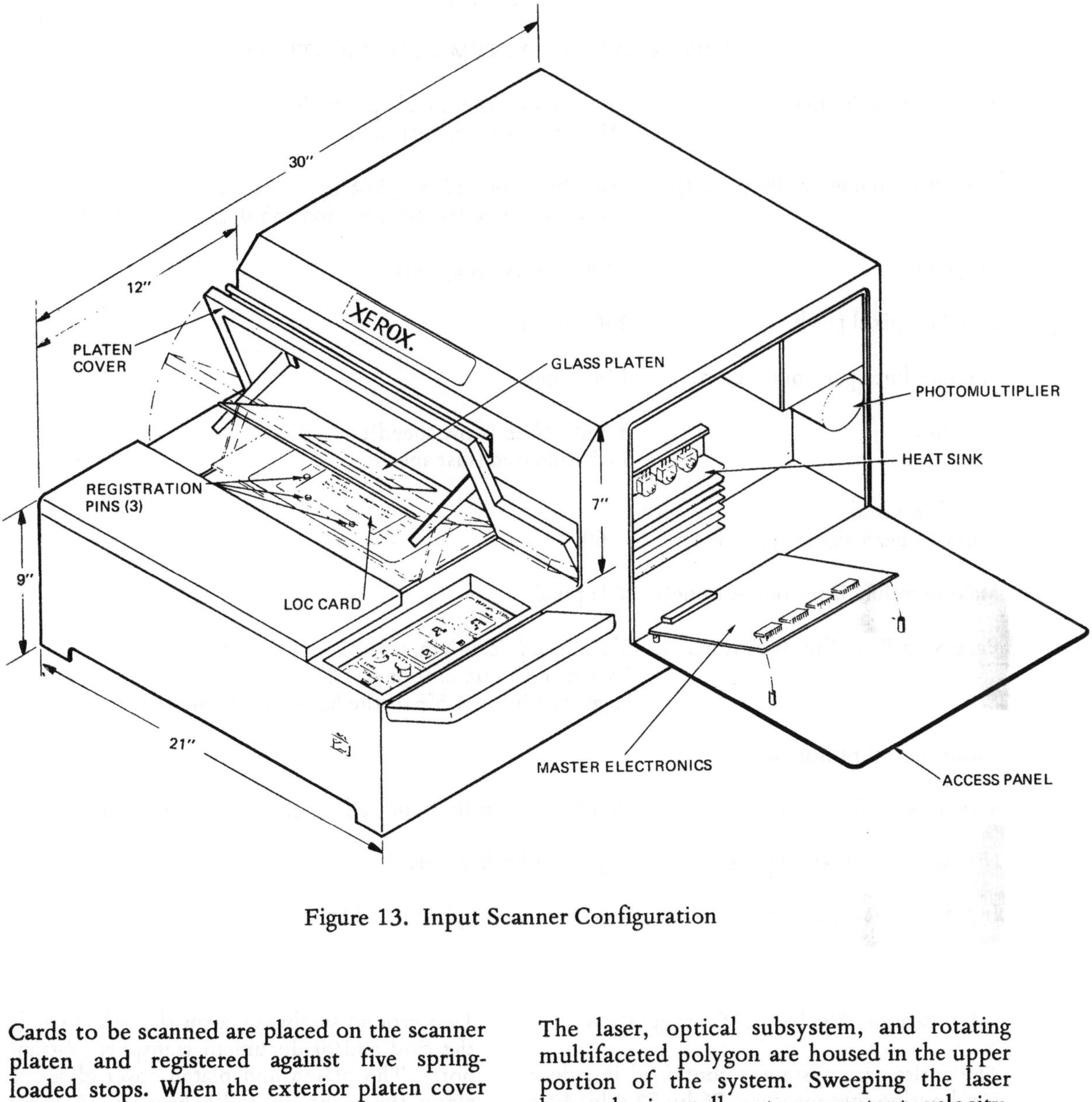

Figure 13. Input Scanner Configuration

Cards to be scanned are placed on the scanner platen and registered against five spring-loaded stops. When the exterior platen cover is closed, a glass plate is automatically lowered to cover the card. This plate and supporting structure are transported with the card into the scanning beam to keep the card flat and registered. In its primary operating mode, the card is transported at a constant velocity of 1 inch/second and returns at 2 inches/second.

The laser, optical subsystem, and rotating multifaceted polygon are housed in the upper portion of the system. Sweeping the laser beam horizontally at a constant velocity, repetitive scans are made as the card or document being scanned is transported at constant velocity in the line-to-line direction. Laser flux reflected from the document being scanned is incident upon a photomultiplier (PMT) detector, thresholded and further conditioned. Specifications for the system are listed in Table 1.

TABLE 1

CARD SCANNER SYSTEM SPECIFICATIONS

Line & pixel density:	18.9 lines & 18.9 pixels per millimeter (480 lines & pixels per inch)
Spot size (diameter 50% points):	Scan direction: 52.83 (2.08 by 103 in.); Cross scan (line to line direction: 55.88μ (2.2 by 10.3 in.)
Polygon:	2400 rpm (slow speed)
Number pixel per line:	2400 pixels
Number lines per library card:	1440 lines
Bit rate:	3.8 Mbit/sec (slow speed); 11.5 Mbit/sec (fast speed)
Signal-to-noise (peak-to-peak signal/rms noise):	$\geqslant$ 60
Maximum line jitter (end-of-line):	$\leqslant$ ¼ pixel
Page Scan Rate (linear motor):	Slow speed: 2.54 cm/sec $\leqslant$ 1% Fast speed: 5.08 cm/sec $\leqslant$ 1% Page scan jitter $\leqslant$ 5% of the laser line-to-line spacing
Number of threshold levels:	$\geqslant$ 64
Max. electronic margin:	Top 18.29 mm (0.72 in); all other sides 9.21 mm (0.3625 in)
Number of margin settings:	Top 32; all other sides 16
Input electrical power:	$\leqslant$ 345 watts (single phase 115V ac)

4.1.1 Scanner Mechanical Configuration

The entire scan system is contained in a desk top housing that measures 76 by 53.4 by 40.6 cm (L x W x H). The system is inclusive of the laser scan subsystem, Y-transport, detection subsystem, electronics, and all power supplies.

Cards to be digitized are registered against five spring-loaded stops on the platen. A double cover interlock permits BRH requirements to be met. When the outer mechanical cover is raised, the glass platen is raised with it, permitting the document being scanned to be inserted. When the outer cover is closed, the glass depresses the registration stops, flattens the cards, and holds the material in place during scanning.

Two ground steelways form the reference for the card and/or document transport (Y-transport). The platen and platen cover glass ride along the ways on roller bearings. A three-bearing cluster, with one spring-loaded assembly, captures the rail. Except for the bearings, the Y-transport is frictionless. There are no other mechanical components, such as lead screws, stepper motors, etc. A coil attached to the document transport surrounds the center pole of the dc motor. Three permanent magnets provide the field and, with 1 ampere of current applied to the coil, the platen is accelerated at approximately 340 cm/sec^2 reaching scanning speed in nearly 7 ms or within a linear distance of less than 0.2 mm. A linear encoder attached to the platen assembly is utilized as a reference for the Y-transport servo.

4.1.2 Scanner Optics

The upper portion of the system, shown in Figure 13, contains the optical scanning system and laser. The HeNe laser and two folding mirrors are mounted on a separate plate. Prealigned and pinned in position, this subassembly is field replaceable. The laser beam reflected from a multifaceted polygon scans across the card or document mounted on the platen. Laser flux reflected from the document enters a light baffle assembly mounted beneath the optics baseplate. A single photomultiplier detector "stares" at the scan line and converts changes in reflectance to threshold video signals.

The optics assembly is a classical post-polygon objective lens system. A schematic of the optical subassembly is shown in Figure 14. Laser relay elements L1, L2, and L3 are utilized to shape the beam prior to scanning. Cylindrical lens element L2 produces a collimated beam from the intermediate waist formed by L1. Collimated in the scan direction, the system is made insensitive to polygon facet radial variations that cause time base inaccuracies and focus errors. Lens L3, having power in only the cross scan (i.e., line-to-line) direction, relays the waist produced by L1 near the polygon surface. The laser beam shape at the polygon is anamorphic, having a waist radii of 60.8μ (cross scan focus) by 1800

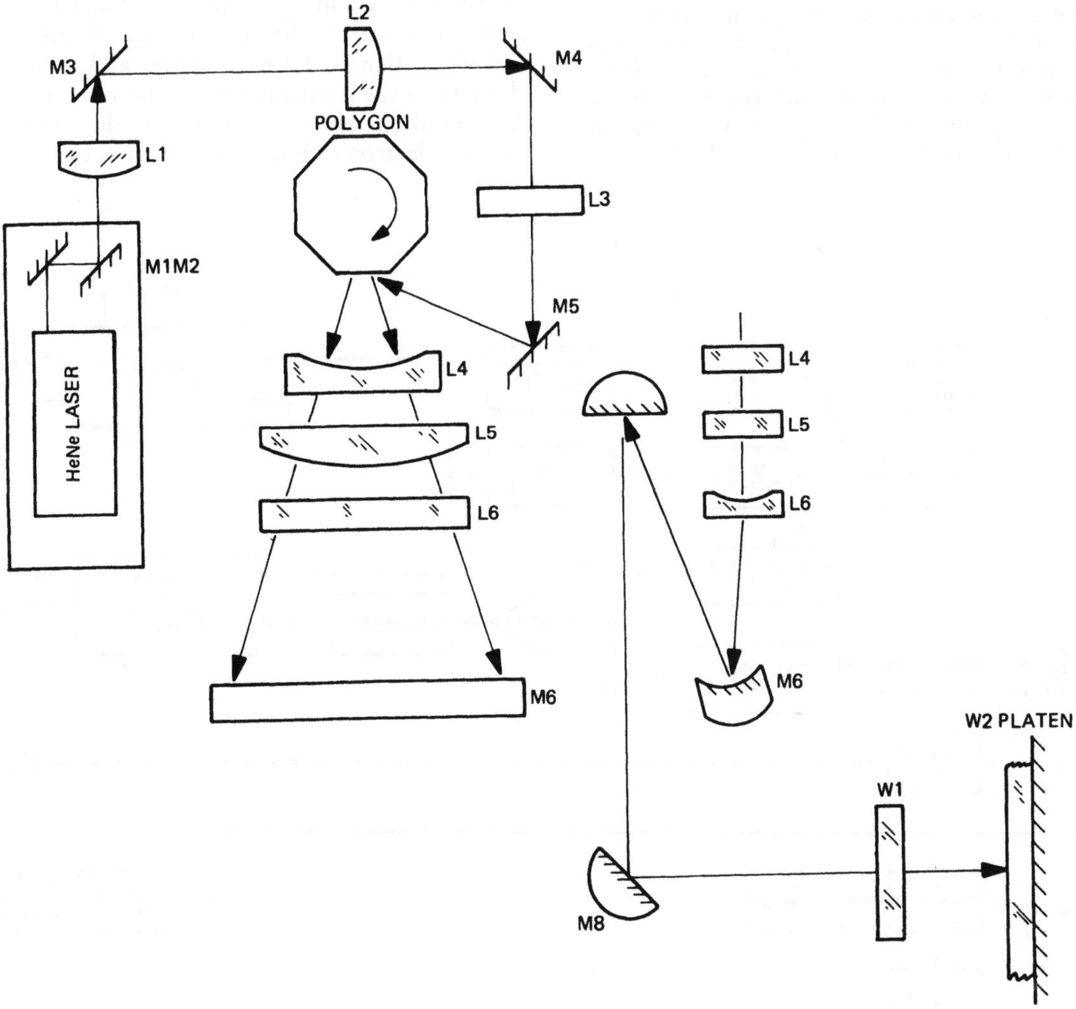

Figure 14. Optical Schematic

(collimated scan plane). The twelve-faceted polygon with facets approximately 2.38 cm across ensures that the beam is untruncated across a 280 mm field. Underfilled, the system provides high radiometric efficiency and untruncated, the optical beam maintains a Gaussian profile throughout.

The post-polygon scan objective lens is comprised of elements L4, L5, L6, and M6. Elements L4 and L5 are cylinders forming an inverted telephoto lens having power in the scan plane. Collimated flux feeding this lens pair is focused into a 53μ spot. Distortion designed into the objective lens linearizes the scan so that the maximum departure from linearity is 50.8μ across a 33 cm field (i.e., 0.015 percent).

Positive and negative elements, L6 and M7, serve a dual function. Firstly, they relay the intermediate laser waist formed near the polygon face to 56μ focus in the cross scan plane. The second function of these two elements is to form a geometric focus of the polygon facet at the focal plane. This latter feature is employed to compensate for polygon pyramidal errors.

Complete optical independence of the scan and cross scan planes permits the system to be easily adjusted to the correct spot size, permits coincidence of scan and cross scan planes to be readily achieved, and permits the focal plane location and correction for polygon pyramidal errors to be easily obtained.

4.1.3 Electronics

The laser input scanner described is interfaced with a PDP-11 computer. It also may be manually operated. A focused laser beam is deflected horizontally by a multifaceted polygon to scan successive lines. As material to be digitized is transported past the reflected laser beam via a linear dc motor, a single PMT detector converts reflected flux into thresholded video signals. Scan detectors and a data clock form the horizontal time base reference. A simplified block diagram of the input scanner electronics is shown in Figure 15.

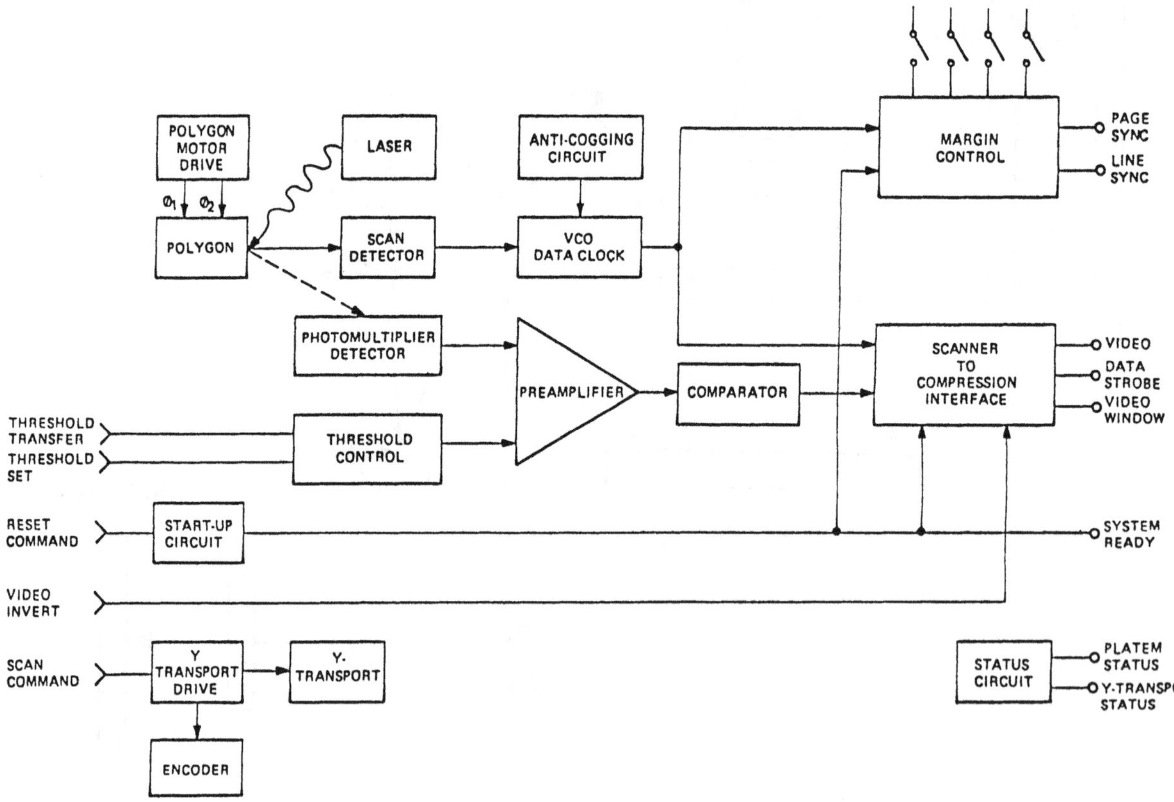

Figure 15. Electronics Block Diagram

In addition to "hunting", inertia loaded synchronous motors exhibit higher frequency speed fluctuations due to motor pole and/or winding slot effects. The major component of this high frequency jitter or "cogging" occurs at twice the excitation frequency and is fixed in both phase and amplitude for any given motor. The major components of this jitter are removed by inserting a correction signal into the control loop through the anticogging circuit. This circuit also corrects for facet-to-facet effects.

Four thumb switches located under a control panel to the right of the platen allow the operator to manually set the top, left, right, and bottom margins. These controls serve to crop the margin and can be set inside the card to delete artificats that are not wanted.

When the Y-transport is in the "home" position, the laser beam scans a reference strip. While scanning this strip, the PMT gain is adjusted by controlling the high voltage. Digitized and clocked, the output is accumulated in a shift register as a parallel 8-bit word and is transferred through the scanner to data compressor circuit.

This document transport or Y-transport, as it is called, is a linear dc motor. Documents to be scanned are placed on a platen and are transported past the laser beam. Drive signals, applied to a coil attached to the platen and surrounded by magnets, provide the force to move the platen document assembly. The Y-transport velocity is referenced to a crystal oscillator running at 5.12 MHz.

4.2 DEMAND OPTICAL DISK

At the inception of the DEMAND program, no optical disk media or drives were available to Xerox from vendors. The use of magnetic storage material for the digitized card images was impractical. Xerox had optical disk technology research work under way. It was decided to develop the optical disk media, the equipment to master the media with registration data and card data, and a read only optical disk drive for use at LOC. This section describes this equipment.

The LOC DEMAND System requires large amounts of digital data storage. The 5,500,000 cards require between 1012 and 1013 bits of data storage with long storage life (e.g., 10 to 50 years minimum).

Figure 16. Optical Disk

The storage medium shown in Figure 16 is a thin film of tellurium evaporated on a glass substrate. Bits of information are actually "holes" melted in the tellurium film. The writing process occurs when a write laser beam focuses on the thin film causing energy to be absorbed at a pinpoint (Figure 17). A hole (i.e., an area without tellurium) develops at a pinpoint on the second surface. Patterns of 0.4 mm diameter holes are found by strobing the laser light along concentric tracks. The tracks are divided into sectors of data along with error coding information based on standard magnetic disk system schemes. The read laser beam operates at a power level substantially below the ablation threshold of the media so that holes may be distinguished from non-holes without affecting the tellurium film integrity.

Indexing information is recorded on the optical media on a Mastering Station at Xerox. Since a read/write capability was not needed, the Mastering Station serves two functions. First, a blank (unrecorded) optical disk must be formatted to identify its track and sector locations and to identify areas that are unsuitable for data recording. This task

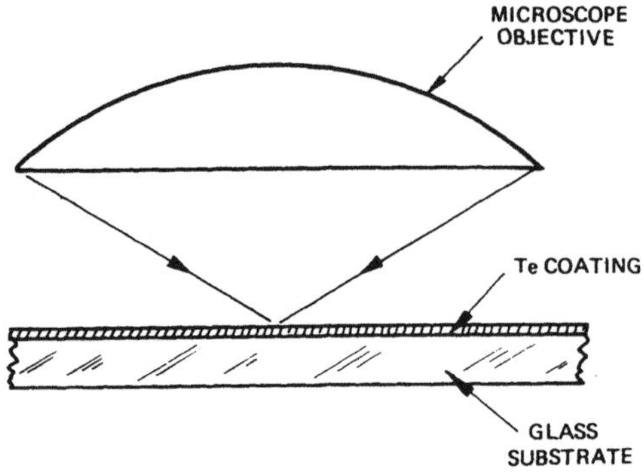

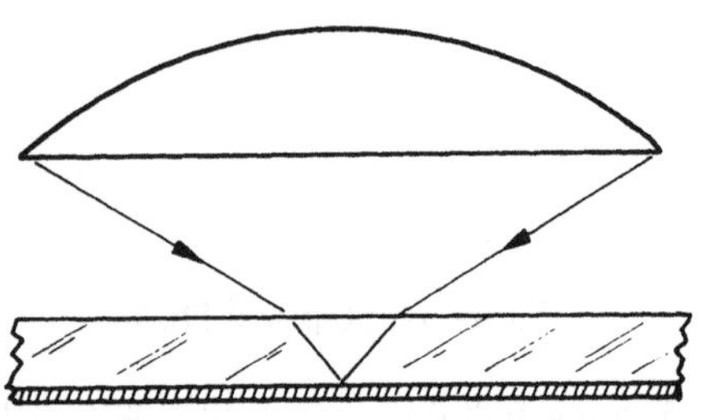

Figure 17. First Surface and Second Surface Recording Configuration

requires a highly precise and mechanically stable mechanism such as that shown in Figure 18. Once formatted, the positions of the optical system for data recording can be accomplished by a less precise mechanism by "track following" the prerecorded information. The Mastering Station serves also as the data recording mechanism upon which data are copied from magnetic disks onto optical disks.

The DEMAND System read only optical disk is pictured in Figure 19. The specifications for the optical disk drive unit are given in Table 2.

4.4 DEMAND SYSTEM PRINTER

LOC has many library cards containing Kanji characters. Because of this, LOC required a system with 480 pixel per inch resolution. To meet this requirement, XSIS made major modifications to a Xerox 9700 Electronic Printer to increase its resolution from 300 to 480 pixels per inch. In addition, XSIS made modifications to the 9700 printer controller electronics to accept the large number of bits of data in each image.

The resolution modification involved replacement of the Raster Output Scanner (ROS) in the 9700. The ROS contains the laser illuminator for the imaging optics, the electro-optical modulator which modulates the digital data on the illuminator light beam and supporting electronics. The DEMAND ROS has:

- o A smaller spot size
- o Wider bandwidth
- o A common illuminator with the standard 9700

These features are required to properly place the high resolution digital images on the light sensitive drum.

The ROS used on the DEMAND program was developed by Xerox on another program and adapted to the DEMAND program needs.

Figure 12 shows the DEMAND program printer. While the unit has the same appearance as a standard 9700, they are very different. The DEMAND printer has the capability of accepting six card images in uncompressed format and delivering a printed image at the printer's rated speed. This capability required the addition of memory and other hardware.

Detail descriptions of the operation of the 9700 are available in the standard Xerox literature. Detail descriptions of the implementation of the special imaginal capabilities are not released at the writing of this paper.

5. TECHNOLOGICAL ADVANCES

Document Delivery System technology has advanced considerably since the beginning of the DEMAND system program. At that time, all major system elements including optical disk, high resolution scanner, and high resolution electronic printers were unavailable in any form. Thus, the DEMAND program was a major development effort. Considerable technological progress has been made since the DEMAND program started.

TABLE 2

DEMAND OPTICAL DISK UNIT SPECIFICATIONS

POWER REQUIREMENT

Frequency (Hz) Single Phase	Voltage
60 Hz	100/110/115/120
50 Hz	200/220/230

Conversion from 60 Hz to 50 Hz may be accomplished in the field

Frequency Tolerance: ±1%

DC power: Not applicable. Power supply is internal.

ENVIRONMENTAL CHARACTERISTICS

	Operating	Shipping/Storage
Temperature	60° to 100° F	-40° to 150° F
Temperature Gradient	20° F/hour	40° F/hour
Temperature Shock	3° F/min	12° F/hour
Maximum Δ T Rise	25° F	–
Relative Humidity	20 to 80%	5% to 90%
Maximum Wet Bulb	78° F	83° F
Barometric Pressure	30.5 to 20.6 inches Hg	31.0 to 5.5 inches Hg
Altitude	-500 to 10,000 ft	-1000 to 40,000 ft
Operating Environment	Maximum acceleration of 0.6 g at 10 to 100 Hz	
Shipping/Storage	Maximum shock of 5 g for 5 ms max vibration is 1 g from 10 to 1000 Hz	
Heat Dissipation Unrestricted air flow to 30 cfm to front of unit with unrestricted exit at rear of unit	2850 Btu/hour average	
Contamination Ambient	Federal Standard 209B Class 1 x 10^{6}	
Acoustic Noise Level	The maximum noise criteria (NC) level generated in the disk drive will not exceed NC60. All noise level measurements are established at a height of 60 inches above the floor and minimum distance of three feet from each side of the machine.	

RELIABILITY/MAINTAINABILITY

Field Repairable	Yes
MTTR	0.5 hour
MTBF	7000 hours
Seek Error Rate	Less than 1 in 10^{6}
Duty Cycle	100%
Preventive Maintenance	Not required
Soft Error Rate (recoverable)	1 x 10^{10} bits transferred
Hard Error Rate	1 x 10^{11} bits transferred
Pack Life	10 Years
Useful Life (Unit)	40,000 hours or 8 years whichever occurs first

TABLE 2 (continued)

DEMAND OPTICAL DISK UNIT SPECIFICATIONS

SAFETY STANDARD REQUIREMENTS

The unit will have the approval of the following safety standards:

- U.L.
- C.S.A.
- B.S.I.
- R.F.I. Standards of the German Government
- V.D.E.

Figure 18. Optical Disk Mastering Station

Figure 19. Library of Congress
Optical Disk Playback Unit

1983 was a very productive year for Document Delivery System product announcements. The following portion of the paper deals with these announcements and speculates on the future.

5.1 SCANNING EQUIPMENT

Since the DEMAND program started, Xerox Printing System Division has announced the high resolution Xerox Model 150 Graphic Input Scanner. This device is compatible with the Xerox 9700, 300 by 300 pixels/inch printing capability. The unit scans all forms of art up to approximately 8½ by 11 inches.

For the engineering documentation market, Versatec, a Xerox Company, has announced the Aperture Card Raster Input Scanner (ACRIS). The ACRIS scans documents and produces raster images encoded at 200 dots per inch.

Companies such as Terminal Data Corporation (TDC), Systems Group of Colorado, and ECRM have scanners available for high volume, moderate resolution large documents, and other special needs.

The laser technology used in the DEMAND program has been supplemented by CCD technology. At present, CCDs are limited in resolution but new devices with increased number of elements are being developed. Until the larger units are ready, optical butting of CCD is being used.

5.2 DISPLAYS

For a systems house, it is still common today to be forced to design a high resolution workstation. However, available standard equipment is better today than in 1978. Examples are Terminal Data Corporation's TDC 2030 and 3010 product lines of high resolution (up to 2000 lines) video terminals.

While video technology is progressing more slowly than CCD technology, the demand for high resolution displays will bring along more improvements in the next two to three years.

5.3 DATA COMPRESSION TECHNOLOGY

The largest aid to this technology will be standardization. Even without enforced standards for all parts of the graphics industry, the integrated circuit manufacturers have compressors and expander chips under development.

5.4 OPTICAL DISKS

In the latter part of 1983, both the Optimem Division of Shugart, a Xerox Company, and Storage Technology Corporation (STC) announced their first optical disk products. The units are quite different and aimed at different markets. The Optimem unit is aimed at the end user and OEM market for low cost drives. The STC optical disk product is designed for the medium to large IBM mainframe user. Table 3 is a comparison of the features of the units.

Juke boxes are being proposed as special fixed price items for the new optical disks. With the jukeboxes there are a wide variety of applications for document delivery systems not possible before.

TABLE 3

STC/OPTIMEM PARAMETERS

Item Description	Parameters: STC	Parameters: Optimem
a. Size		
(1) Disk drive	52" W by 55" H by 36" D	19" W by 7" H by 26" D
(2) Disk	14"	12"
(3) Disk controller	45" W by 71" H by 32" D	19" W by 12" H by 26" D
b. Disk capacity	4 x 10^9 bytes	1 x 10^9 bytes
c. Transfer rate	3 & 1.5 MB/s	0.55 MB/s
d. Average access	80.3 ms	160 ms
f. System compatibility	Primary, IBM; secondary general with emulator added	General

Given the early state of the optical disk products, there is plenty of potential for reduction in both the price of the drives and the media. Larger memories are also quite predictable.

Erasable optical disk is under development but still in the future.

5.5 ELECTRONIC PRINTING SYSTEMS

The DEMAND and CARDS systems were designed during the early days of the Xerox 9700. Since then, the 300 pixel per inch Xerox 9700 has become a very popular product. Only recently have special printing products with higher resolution been released. These products and market trends indicate some market accommodation for higher resolution products.

For the document delivery market with needs for larger than 8½ by 11 paper documents, the market has been satisfied to date by plotters such as those manufactured by Versatec. XSIS is offering large engineering document delivery systems implemented with either Versatec plotters or electronic printers based on the high volume Xerox 2080 and Xerox 7080 engineering document reproduction equipment.

The Xerox 2080 and the more recently announced Xerox 7080 copiers are ideal for high volume systems where cut sheet D size plain paper in high volume copies is required. The 7080 even has folding, stitching, and stapling equipment.

5.6 CONCLUSION

While DEMAND is the first of the operational demand printing systems based on optical disks, there will be many more systems operational in the next two to five years.

6. ACKNOWLEDGMENTS

This paper had major contributions by the Document Delivery Systems technology team at XSIS. Major contributors were Mr. H. D. Wolpert, Program Manager for the Laser Scanner; Mr. Richard Krueger, XSIS Senior Marketing Specialist; and Mr. William Peters, DEMAND Program Manager.

7. BIOGRAPHY

Mr. Lee F. Wilbraham is Manager, Advanced Systems Concepts and Planning for Xerox Special Information Systems (XSIS). XSIS is the Systems Division of Xerox Corporation. He is responsible for product planning, business planning, customer system study program direction, system design and cost estimating, and proposal preparation.

Mr. Wilbraham is a graduate of Purdue University and Denison University in Electrical Engineering and Physics. His graduate work in Business Administration was done at California State University at Northridge.

Mr. Wilbraham's career includes approximately 25 years in the computer and engineering systems business. Before joining Xerox, he was with Litton Industries, Beckman Systems Division, Scientific Data Systems, and Interstate Electronics Corporation.

9780898381573